D0186245

Practice Papers for SQA Exams

CfE Higher

Geography Practice Papers

© 2015 Leckie & Leckie Ltd

001/22042015

10 9 8 7 6 5 4 3 2 1

ISBN 9780007590995

Published by
Leckie & Leckie Ltd
An imprint of HarperCollins*Publishers*
Westerhill Road, Bishopbriggs, Glasgow, G64 2QT
T: 0844 576 8126   F: 0844 576 8131
leckieandleckie@harpercollins.co.uk
www.leckieandleckie.co.uk

Publisher: Katherine Wilkinson
Project manager: Craig Balfour

Special thanks to
Donna Cole (copy edit)
Louise Robb (proofread)
Ruth Hall (proofread)
QBS (layout & illustration)
Ink Tank (cover)

Printed in Italy by Grafica Veneta S.p.A

A CIP Catalogue record for this book is available from the British Library.

Acknowledgements

This product uses map data licensed from Ordnance Survey © Crown copyright and database rights (2014) Ordnance Survey (100018598). Pages 22, 36–37, 51 and 64–67.

The following were adapted from SQA questions with permission, Copyright © Scottish Qualifications Authority: Exam A questions 3, 4 and 7; Exam B questions 1, 2 and 4; Exam C questions 1 and 7; Exam D questions 3, 4 and 9.

Illustrations © HarperCollins Publishers

P13 DrimaFilm / Shutterstock.com; P67 Geraint Thomas © ADRIAN DENNIS / Staff / Getty Images

All other images © Shutterstock.com

Whilst every effort has been made to trace the copyright holders, in cases where this has been unsuccessful, or if any have inadvertently been overlooked, the Publishers would gladly receive any information enabling them to rectify any error or omission at the first opportunity.

Introduction

Layout of the book

This book contains four practice exam papers, which replicate the Higher Geography SQA exam as much as possible. The layout, structure, question style and levels are similar to those of the final exam. Using practice papers is an ideal way to practice prior to the final exam and will make you more familiar with the layout of the exam.

Answers are provided for each of the four practice papers. Each answer includes comments that meet the standard to achieve marks. There are also tips on important exam techniques that will help you gain valuable marks and avoid common mistakes.

How to use this book

There are many ways to use this book. You may want to dip in and out of sections as you cover them in class or before important assessments. Or you may wish to tackle a full paper.

Use the topic index to help you locate specific questions if you are looking for questions on a particular topic. You will find practice questions for each topic. This lets you focus on particular areas of difficulty or recap during and at the end of topics as you cover them in class. To begin with, do not worry about consulting the marking instructions as these often jog our memories and also show the level required and content that the examiner is expecting in your answer.

It is worth completing at least one of the papers as preparation for the final exam. At the outset of your course you will not be able to do this, but as you near the final exam you should have covered enough content to tackle an entire paper. At the start you might want to break the exam down into sections and consult your notes but as you near the end of your revision you should consider attempting an exam without your notes and within the time permitted.

Overview of Higher Geography course

You will complete unit assessments throughout the year in Higher Geography. You must pass all of these as well as the course assignment (an added value unit). There are a number of ways to achieve these unit assessments, for example you may sit a test, complete a written report or give a presentation. The purpose of the internal assessments is for you to demonstrate that you meet the standard required to pass the course.

The final course award you receive will be based on the marks you receive in the two assessed components:

*   The question paper – 60 marks

*   The assignment – 30 marks

This gives a total mark out of 90. This mark will be converted into a percentage and then a grade.

Grades will normally be awarded A–D on the following basis (note: this can vary slightly each year):

- Grade A = 70–100%

- Grade B = 60–69%

- Grade C = 50–59%

- Grade D = 45–49%

The assignment

You will be given more details about the assignment by your teacher during the academic session. It is important that you are well-prepared for the assignment as it makes up a portion of your final grade.

The assignment will assess your skills and knowledge of a particular geographic topic or issue of your choice. Discuss this with your teacher as you will need to spend a lot of time preparing. The assignment is completed during an independent write-up prior to the final exam. You will be given 1 hour and 30 minutes to do this.

To achieve a good mark in the assignment you must do the following:

- Identify a suitable topic.

- Carry out research – this may include independent field work or field work as part of a Geography trip.

- Demonstrate an awareness of suitable methods for the collection of geographic data. You should also comment on the suitability and reliability of these methods.

- Process information you have gathered using a range of tables, graphs etc.

- Connect the findings you have made to subject knowledge.

- Analyse the information you have gathered.

- Reach a conclusion based on the evidence you have collected.

The exam

You will sit one final exam for Higher Geography. The exam lasts 2 hours and 15 minutes and is worth 60 marks. The exam is made up of four sections.

- Section 1 – Physical Environments, which is worth 15 marks. You may be asked about some or all of the physical topics in this section. **You must answer every question in this section.**

- Section 2 – Human Environments, which is worth 15 marks. You may be asked about some or all of the human topics in this section. **You must answer every question in this section.**

- Section 3 – Global Issues, which is worth 20 marks. There are five questions in this section, covering a range of topics. You will have studied at least two of these topics. All of the questions are worth 10 marks. **You must answer two of these questions.**

- Section 4 – Application of Geographical Skills, which is worth 10 marks. In this section you will be tested on your mapping skills and use of numerical and graphical information. For a good mark you must use the OS map and all of the information given in the question. **You must answer this question.**

Examination advice

The obvious advice is really important and applies to all of your exams:

- Make sure you know when your final exam is.

- Arrive early for the exam.

- Bring blue or black ink pens and pencils for diagrams.

- Write legibly so that your answers are understood. Consider leaving lines between paragraphs and a short section between each question, so that you have space should you wish to return to this question.

- Make sure you are well aware of the exam structure: How many questions you should answer in each section? How long the exam lasts?

The following apply to the answers you give. These tips are crucial as you may be losing marks without realising it. If you are not sure what any of them mean, look at the tips in the marking instructions section at the back of this book for more detail. Alternatively ask your teacher and/or classmates.

- Read each question very carefully. Identify the command words (see the next section) and respond accordingly. Make sure you are totally sure what you are being asked before you start writing.

- Plan your answer by writing down key words. Add reminders of important things to include in your answer and cross these off as you go. Do not make this too long, but spending a short time thinking out your answer will definitely pay off.

- In longer questions and/or questions where you are asked to do two things (e.g. social and environmental impacts), use subheadings and paragraphs to provide structure. This makes your answers easier to follow and will ensure that you cover each area of the question (there are penalties if you do not).

- Give full explanations. In Higher you must explain and give reasons. For example, when covering physical processes such as abrasion and hydraulic action you must explain in full how these processes operate. Linking words and phrases such as 'this means that', 'therefore' and 'because' will force you to conclude points. Make sure you link these points to the question.

- Watch your time. The exam is 2 hours and 15 minutes. You must finish the exam to do well. Pace yourself and move through each section or you will not complete the exam. As a rule of thumb, you have a little over 2 minutes per mark (i.e. a 5-mark question should take you about 11 minutes). Make sure you look at the number of marks available and vary the length of your answer accordingly.

- Where appropriate, give examples from the case studies you have learned. Remembering place names, events and statistics will be expected of very good candidates. Be careful though – learn too many and you will struggle to include them and may get confused.

- Use geographic terminology throughout your answers. For example, when areas of the inner city are redeveloped this leads to growth as new companies come into the area, creating jobs and improving services, and better candidates will refer to this process as the 'multiplier effect'.

- Practice drawing annotated diagrams that you can reproduce quickly in the exam. Marks will not be given for your artistic ability but for the information that you add as labels.

There are a number of common errors to avoid during the exam:

- Answering more than you need to. Do not answer more than two questions in Section 3 of the exam paper. If a question has an option, only write about one option. Similarly, if you write more for a 4-mark question than a 7-mark question you are going to lose marks somewhere.

- Not answering the whole question. If you are asked for solutions and the effectiveness of these you will forfeit some marks if you do not talk about both parts of the question.

- Not referring to sources in the Section 4 question. You will need to make good use of the map and any numerical data given to you.

- Listing answers. You will see from the marking instructions that you need to write in fully developed sentences to achieve marks.

- Not completing the exam. Use this book to practice your timing. Your best chance of doing well is to pace yourself evenly through the paper and to complete every question.

- Giving irrelevant information in answers. You must stick to the question. For example, if you are asked about the impact of global warming you should not write about the causes.

- Writing vague and/or over-generalised answers. In these types of answers candidates may name a city or country but other than this it appears that they know very little about their case study.

- Reversals. This is where you explain the exact opposite in your answer. This will not gain additional credit. For example, if you were asked about migration you may write about push factors such as lack of jobs and poor health-care. You must be careful that when discussing pull factors you simply do not just write about the opposite, e.g. employment opportunities and lots of doctors in hospitals.

Command words

In the exam a number of command words will appear. These may be in **bold** within the question to draw your attention to them. They indicate how you should approach each question and demand different approaches. Take your time to reflect on the table below and look at the examples within this book and the corresponding marking instructions.

Each of the command words requires you to write in fully developed sentences with examples and explanations. If you write in a descriptive manner or you do not develop your answers you will be limited to a maximum of half the marks, so it is important to master this skill throughout the year.

These command words are designed to test higher order skills.

Command word	Meaning/Explanation
Explain/suggest reasons	Think about why something has happened. What are the reasons and/or processes behind an action or outcome? For example, if you are asked to explain the formation of a corrie, it is important that you are able to demonstrate an understanding of the processes and conditions involved.
Analyse	In this type of question you should give an account of the relationship between factors. Links will be important. For example, you may be asked to analyse the various properties in the formation of a soil.
Evaluate	In this question you are being asked if a solution/management strategy etc. has been a success or a failure. You should briefly explain the strategy before assessing the impact it has had. In this question type it is important to back up your answer with evidence and to have a variety of ideas.
Account for	Give reasons – this may often come from a source given in the exam paper. For example, you may be asked to account for the rise in global temperatures with reference to physical and human factors.
Discuss	Give the key features of different viewpoints or consider the impact of change. For example, you may be asked to discuss the possible impacts of trade inequalities around the world.
To what extent	Similar to 'evaluate' in certain contexts. Comment, with evidence, on the impact of a strategy. For example, to what extent have the methods used to try and control the spread of malaria been effective?

Topic Index

TOPIC	QUESTION FOCUS	EXAM A	EXAM B	EXAM C	EXAM D
Physical Environments					
Atmosphere					
Global heat budget	Distribution of solar energy				Q3
Redistribution of energy by atmospheric and oceanic circulation	Ocean and atmospheric currents		Q1		
Cause, characteristics and impact on the Intertropical Convergence Zone	Impact of ITCZ on rainfall in West Africa	Q2		Q2	
Hydrosphere					
Hydrological cycle within a drainage basin	Movement of water within a drainage basin			Q1	
Interpretation of hydrographs	Rural and Urban hydrographs	Q3			
Lithosphere					
Formation of erosion and depositional features in glaciated and coastal landscapes	Features of coastal erosion: Headland and Bay formation	Q1			
	Formation of features of glacial deposition				Q2
Rural land use conflicts and their management related to glaciated and coastal landscapes	Impact of tourism		Q3		
	Land use conflict and solutions / management			Q3	
Biosphere					
Properties and formation processes of podzol, brown earth and gley soils	Soil properties		Q2		
	Formation of gley soil				Q1
Human Environments					
Population					
Methods and problems of data collection	Why countries collect population data		Q5		
	Challenges collecting census data				Q6
Consequences of population structure	Impact of ageing population	Q4			
	Impact of a rapidly growing population			Q5	
Causes and impacts of forced and voluntary migration	Causes of voluntary migration				Q5

Rural					
The impact and management of rural land degradation related to a rainforest or semi-arid environment	Impact of rural land degradation on people and the environment			Q6	
	Management of rural land degradation		Q4		
Urban					
The need for management of recent urban change (housing and transport) in a developed and developing world city	Problems caused by rapid population growth in developing countries in urban areas	Q4a		Q4a	
The management strategies employed	Traffic management strategies	Q5			
	Solutions to housing problems in the inner city in developed countries		Q6		
	Solutions to housing problems in developing countries				Q4a
The impact of management strategies	Effectiveness of traffic management strategies	Q5			
	Effectiveness of solutions in developing world cities		Q4b		
	Effectiveness of solutions to housing problems in developing countries				Q4b
Global Issues					
River Basin Management					
Physical characteristics of a selected river basin	Physical characteristics of a river basin	Q6a	Q7a		
The need for water management	Need for water management	Q6b			
Selection and development of sites	Physical / human factors affecting the site of a dam		Q7a		Q7a
Consequences of water control projects	Benefits / problems of a water management projects		Q7b	Q7b	
Development and Health					
Validity of development indicators	Composite indicators of development			Q8a	
Differences in levels of development between developing countries	Differing levels of development between developing countries	Q8b		Q8a	

Specification content	Content detail				
A water-related disease: causes, impact and management	Human and environmental conditions that put people at risk of spreading disease	Q7b			
	Social and economic impact of disease			Q8a	
	Management of disease			Q8b	
Primary health care strategies	Suitability of primary health-care strategies	Q7a			Q8b
Global Climate Change					
Physical and human causes	Physical factors responsible for changes in global temperatures	Q8a			
	Human factors responsible for changes in global temperatures				Q9a
Local and global effects	Impact of global warming		Q9a	Q9a	
Management strategies and their limitations	Measures to control problems related to climate change	Q8b			
	Success of strategies to combat climate change		Q9b	Q9b	Q9b
Trade, Aid and Geopolitics					
World trade patterns	Patterns of global trade	Q9a			
Causes of inequalities in trade	Reasons for trade inequalities		Q10a	Q10a	
Impact of world trade patterns	Impact of world trade on developing countries	Q9b			Q10a
Aid and other strategies to reduce trade inequalities and their impact	Methods of reducing trade inequalities / impact of these methods	Q10b	Q10b	Q10b	Q10b
Energy					
Global distribution of energy resources	Global distribution of energy resources	Q11a	Q11a	Q11a	Q11a
Reasons for increase in demand for energy in both developed and developing countries	Growth in demand of energy	Q10a			
Effectiveness of renewable and non-renewable approaches to meeting energy demands and suitability within different countries	Effectiveness of non-renewable approaches to meeting energy demands		Q11b		
	Effectiveness of renewable approaches to meeting energy demands	Q11b		Q11b	Q11b
Application of Geographical Skills					
This question will test your ability to handle a range of resources including OS map extracts, tables and graphs		Q11	Q12	Q12	Q12

Practice Papers for SQA Exams

HIGHER
GEOGRAPHY
Exam A

Duration – 2 hours and 15 minutes

Total marks – 60

SECTION 1 – PHYSICAL ENVIRONMENTS – 15 marks

Attempt ALL questions.

SECTION 2 – HUMAN ENVIRONMENTS – 15 marks

Attempt ALL questions.

SECTION 3 – GLOBAL ISSUES – 20 marks

Attempt TWO questions.

SECTION 4 – APPLICATION OF GEOGRAPHICAL SKILLS – 10 marks

Attempt the question.

Credit will be given for appropriately labelled sketch maps and diagrams.

Write your answers clearly in the answer booklet provided. In the answer booklet you must clearly identify the question number you are attempting.

Use **blue** or **black** ink.

Note: The reference maps and diagrams in this paper have been printed in black ink only. No other colours have been used.

Scotland's leading educational publishers

SECTION 1: PHYSICAL ENVIRONMENTS – 15 marks

Attempt ALL questions

Question 1

Headlands and bays are landscape features present in coastal landscapes.

Explain the conditions and processes involved in the formation of a headland and bay. You may wish to use an annotated diagram or diagrams.　　　**5**

Question 2

The ITCZ is a lifeline for the people and environment of West Africa.

Analyse the impact of the ITCZ on the rainfall pattern in West Africa.　　　**4**

Question 3

Diagram Q3 shows two flood hydrographs in the aftermath of a heavy rainstorm.

Account for the differences in discharge between the urban and rural hydrographs.　　　**6**

Diagram Q3: Flood Hydrographs

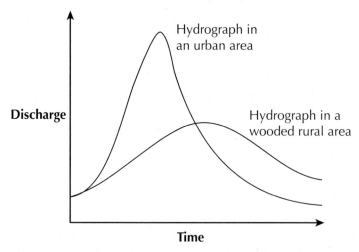

SECTION 2: HUMAN ENVIRONMENTS – 15 marks

Attempt ALL questions

Question 4

Study Diagram Q4

Scotland's population has changed dramatically in the last few decades.

With reference to a country or countries you have studied, **discuss** the impact of the changes shown in diagram Q4 on countries such as Scotland.　　**6**

Diagram Q4: Scotland's population 2012

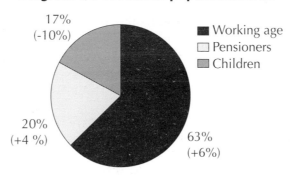

17% (-10%)

20% (+4 %)

63% (+6%)

■ Working age
□ Pensioners
▨ Children

Question 5

Look at Diagram Q5.

For a city in the **developed world** you have studied:

   (a) **Explain** the techniques used to combat traffic congestion; and consider

   (b) **To what extent** these techniques have been effective?　　**9**

Diagram Q5: Traffic measures during Glasgow Commonwealth Games 2014

Traffic concerns were one of the biggest challenges for organisers of the Glasgow 2014 Commonwealth Games. Moving athletes and visitors around the city and ensuring they were on time for events was crucial to the success of the Games. Organisers also had to ensure that the residents and workers of Glasgow were able to go about their day-to-day business.

During the 2014 Commonwealth Games in Glasgow a comprehensive transport plan was put in place to keep competitors, visitors and residents moving. The planning process involved representatives from the local and business community as well as public transport providers and the local authorities.

SECTION 3: GLOBAL ISSUES – 20 marks

Attempt TWO questions

Question 6 – River Basin Management

(a) **Analyse** the physical characteristics of a river basin you have studied.   **5**

(b) Look at Map Q6 and Diagram Q6A and Diagram Q6B.

Referring to Egypt and using the resources provided, **explain** why there is a need for water management.   **5**

Map Q6

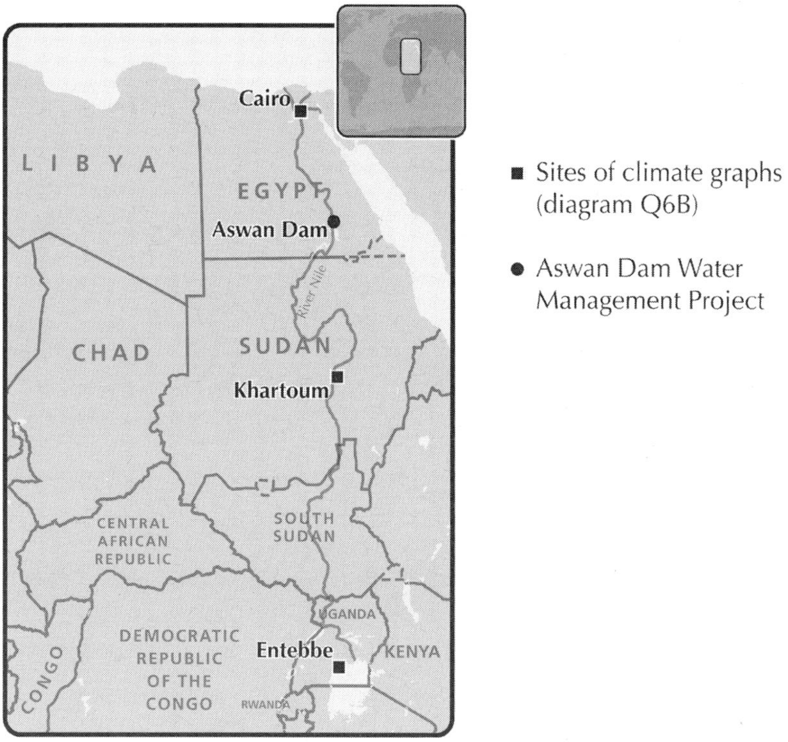

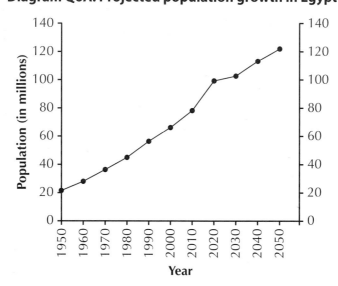

■ Sites of climate graphs (diagram Q6B)

● Aswan Dam Water Management Project

Diagram Q6A: Projected population growth in Egypt

Diagram Q6B: Climate graphs

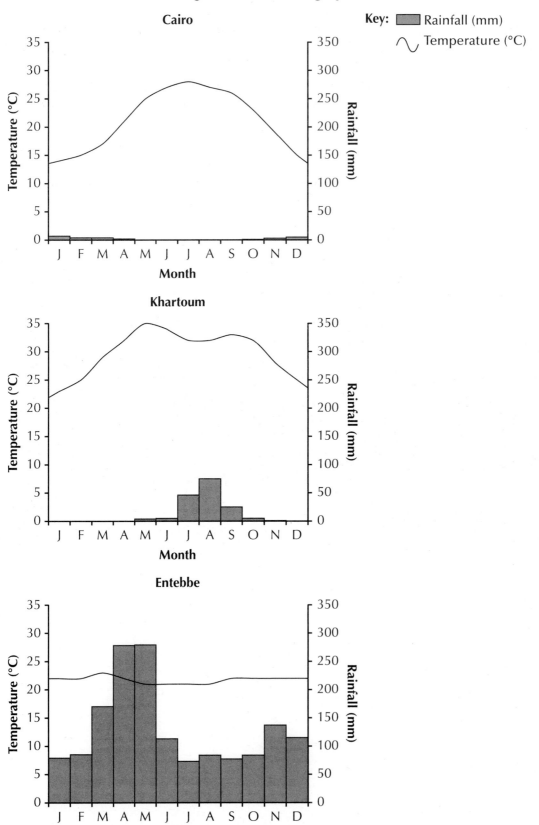

Key: Rainfall (mm)
     Temperature (°C)

Cairo

Khartoum

Entebbe

Question 7 – Development and Health

> *'Resources need to be targeted at improving Primary Health Care if we are ever going to improve the health of people in developing countries.'*
> Aid Worker

(a) Referring to named examples, **explain** why improving health standards through primary health-care strategies is suited to people living in developing countries. **5**

(b) Look at Diagram Q7.

For malaria, or any other water-related disease you have studied, **explain** the human **and** environmental conditions that put people at risk of contracting this disease. **5**

Diagram Q7

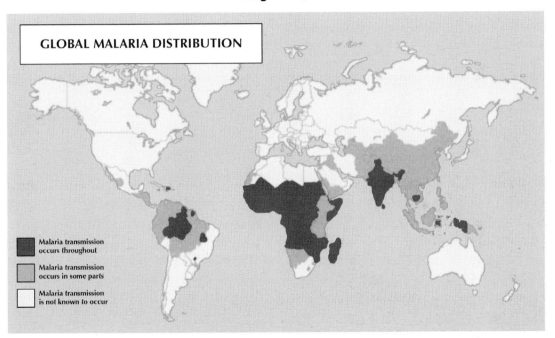

GLOBAL MALARIA DISTRIBUTION

Malaria transmission occurs throughout

Malaria transmission occurs in some parts

Malaria transmission is not known to occur

Question 8 – Global Climate Change

Look at Diagram Q8.

   (a) **Explain** the physical factors that have contributed to changes in global air temperatures.   **5**

   (b) **Explain** measures taken to combat the problems associated with climate change.   **5**

Diagram Q8

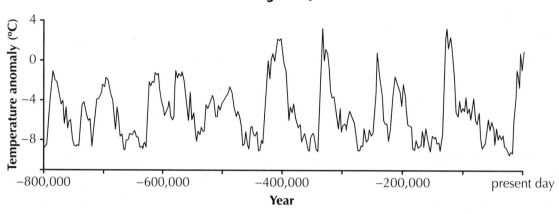

Using data taken from ice cores, scientists are able to construct graphs showing global temperatures over the long-term. The graph shows how temperature varies in comparison to the average global temperature.

Data collected shows that there have been significant ice ages and periods of warming over the last 800,000 years.

Question 9 – Trade, Aid and Geopolitics

Look at Diagram Q9.

(a) Developed countries have traditionally had a larger share of global trade than developing countries. With reference to developed and developing countries, **account for** patterns of global trade. **5**

(b) **Discuss** the impact of world trade patterns on **developing** countries. **5**

Diagram Q9: Export of goods by region (US$ billion)

	Manufactured products	Fuel and mining products	Agricultural products
Europe	4910	812	708
Asia	4566	690	390
North America	1616	408	266
South/Central America	194	297	217
Middle East	276	880	33
Africa	112	397	62

MARKS

Question 10 – Energy

(a) Look at Graph Q10.

Account for the growth in demand for energy in **developing** countries.

4

(b) **Evaluate** the success of non-renewable approaches to meeting global energy demands. You should refer to different countries in your answer.

6

Graph Q10: World energy consumption

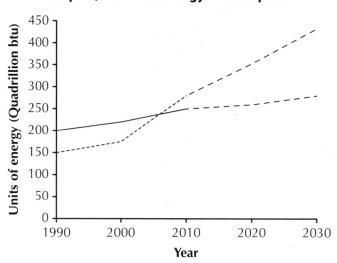

Key: —— Developed countries
----- Developing countries
- – – Projected

SECTION 4: APPLICATION OF GEOGRAPHICAL SKILLS – 10 marks

Attempt the question

Question 11

West Lulworth is a small village on the south coast of England. As well as being popular with tourists, it is a rural community surrounded by many working farms.

The planning proposal below outlines plans for a new holiday park to be constructed in West Lulworth. The plans have been approved but a decision has to be made about which site to locate the holiday park on in the area. Two sites in the area are being considered.

Planning proposal

A plan has been approved in principle for the construction of a holiday park. The park will contain:

a. A camp site with space for 20 pitches (tents)

b. Berths for 10 caravans

c. A shower and laundry block

d. A café and bar

e. A car park for 10 vehicles (parking will also be available next to each caravan)

The investment of £350,000 will be part funded by the regional council, the rural development fund and private holiday firm Flynn Holidays.

Study Map Q11A: OS map; Map Q11B: Dorset coastline; Map Q11C: Proposed sites of new holiday park; Diagram Q11A; Diagram Q11B, Diagram Q11C and Diagram Q11D.

Referring to map evidence and other information from the sources:

(a) **Account for** the large number of visitors who are drawn to West Lulworth each year.   **3**

(b) Decide which site, A or B, is the best site for the proposed holiday park. You must **explain** your choice.   **7**

Map Q11B: Dorset coastline

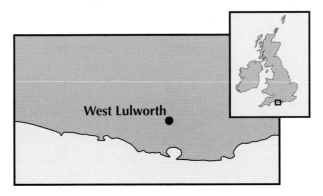

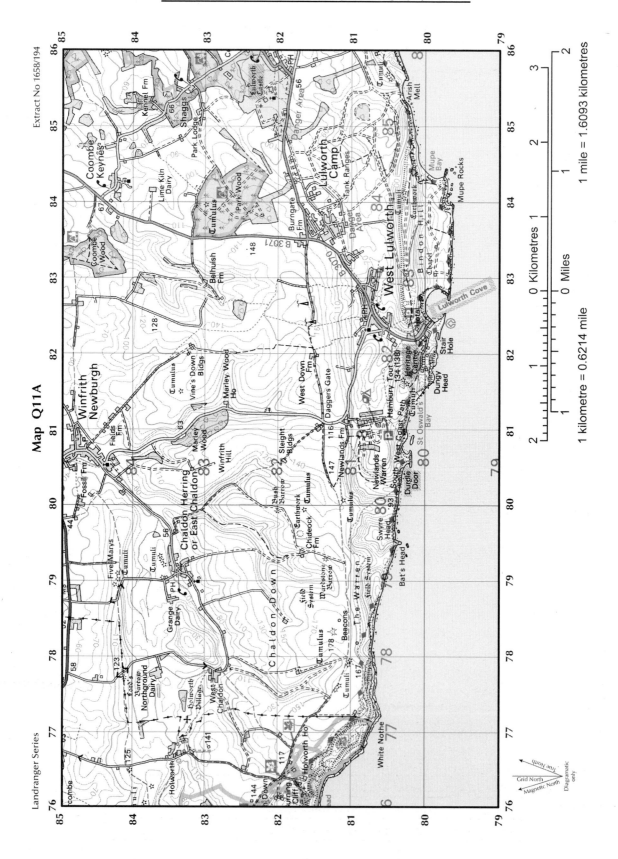

Map Q11A

Extract No 1658/194

Landranger Series

1 mile = 1.6093 kilometres

1 kilometre = 0.6214 mile

True North
Grid North
Magnetic North
Diagramatic only

Map Q11C: Proposed sites of new holiday park

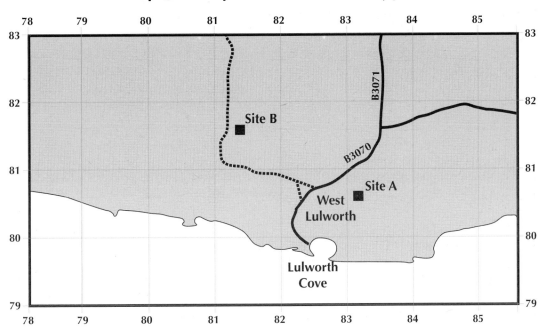

Diagram Q11A: Letter extract, *West Lulworth Weekly Observer,* 6 June

Dear Editor

The lack of accommodation in the local area is putting visitors off coming to West Lulworth or staying in the area longer than a few hours. While young people in other small towns in Dorset are moving into tourist jobs and staying in the area, youngsters in the West Lulworth district are leaving home and moving to nearby towns and cities. It is important for the local economy that the plans for the new holiday park are approved as a matter of urgency.

Yours sincerely

Lisa Hotchkiss

Managing Director

Flynn Holidays

Diagram Q11B: Selected tourism data for Dorset coastline

21 million day visits annually
12% of all jobs in Dorset are tourism related
Annual spend by day visitors £682 million
Tourists spend on second homes £3.3 million

Diagram Q11C: Land use in the county of Dorset

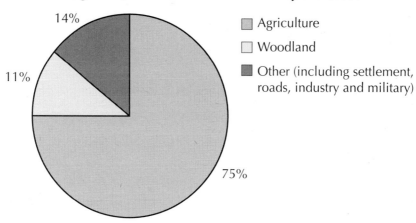

Legend:
- Agriculture
- Woodland
- Other (including settlement, roads, industry and military)

14%
11%
75%

Diagram Q11D: Employment by broad industrial sector

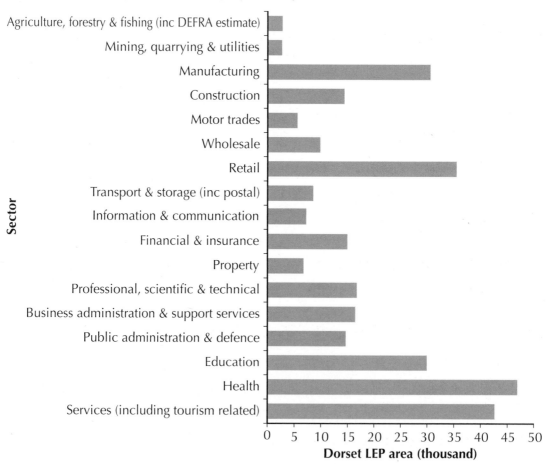

Sector (y-axis) / Dorset LEP area (thousand) (x-axis, 0 to 50)

[END OF PRACTICE QUESTION PAPER]

Practice Papers for SQA Exams

HIGHER
GEOGRAPHY
Exam B

Duration – 2 hours and 15 minutes

Total marks – 60

SECTION 1 – PHYSICAL ENVIRONMENTS – 15 marks

Attempt ALL questions.

SECTION 2 – HUMAN ENVIRONMENTS – 15 marks

Attempt ALL questions.

SECTION 3 – GLOBAL ISSUES – 20 marks

Attempt TWO questions.

SECTION 4 – APPLICATION OF GEOGRAPHICAL SKILLS – 10 marks

Attempt the question.

Credit will be given for appropriately labelled sketch maps and diagrams.

Write your answers clearly in the answer booklet provided. In the answer booklet you must clearly identify the question number you are attempting.

Use **blue** or **black** ink.

Note: The reference maps and diagrams in this paper have been printed in black ink only. No other colours have been used.

Scotland's leading educational publishers

SECTION 1: PHYSICAL ENVIRONMENTS – 15 marks

Attempt ALL questions

Question 1

> *'Energy is transferred from areas of surplus to areas of deficit. Atmospheric and oceanic circulation help to redistribute this energy from areas of surplus towards areas of deficit.'*

Explain how the atmosphere **and** oceans help to maintain the global energy balance.   **6**

Question 2

Look at Diagram Q2.

Account for the differences in the properties of a podzol and brown earth soil.   **5**

Diagram Q2: Selected soil profiles

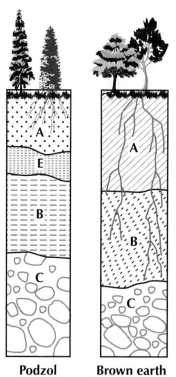

Podzol   Brown earth

Question 3

Discuss the negative effects associated with tourism in a **coastal** landscape you have studied.   **4**

SECTION 2: HUMAN ENVIRONMENTS – 15 marks

Attempt ALL questions

Question 4

> *'Lagos, situated in south-west Nigeria, is one of the world's mega-cities. Rates of crime are high and its residents live in harsh climatic conditions. 66% of residents live in slums. The government estimates the population will soon have reached 25 million, an increase of 10 million in a relatively short period of time. Lagos is on track to become the third largest city in the world.'*
>
> Rapid population growth is one of the biggest challenges facing cities in the **developing world.** In many cities new migrants often end up in shanty towns/favelas.

Diagram Q4: Shanty town population by continent

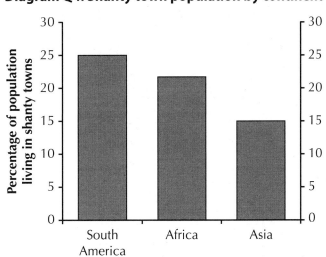

For a city in the developing world you have studied:

(a) **Explain** the main social, economic and environmental problems facing residents and local authorities as a result of rapid population growth.

(b) **Evaluate** the impact of methods used to overcome these problems. **7**

Question 5

With reference to the census and other methods of data collection, **explain** why governments collect population data. **4**

Question 6

With reference to **either** a named rainforest **or** semi-arid area, **explain** the impact of rural land degradation on people **and** the environment. **4**

SECTION 3: GLOBAL ISSUES – 20 marks

Attempt TWO questions

Question 7 – River Basin Management

(a) **Explain** the physical **or** human factors that were important in the selection and development of a site for a named water management project you have studied. **4**

(b) **To what extent** has this water management project been successful? Give **reasons** for your answer. **6**

Question 8 – Development and Health

(a) Look at Diagram Q8.

The Human Development Index (HDI) is a composite indicator of development. Referring to the HDI, or any other composite indicator you have studied, **explain** why it is a useful indicator of level of development. **4**

Diagram Q8

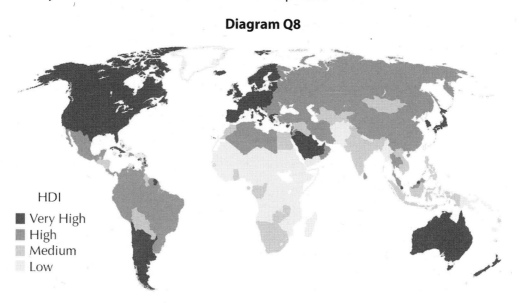

HDI

■ Very High
■ High
■ Medium
■ Low

(b) **Account for** the differences in levels of development between developing countries. You should refer to named examples in your answer. **6**

Question 9 – Global Climate Change

Look at Diagram Q9. It shows the rise in sea level required to flood the lowest point in a range of cities around the world.

(a) **Discuss** the impacts of global warming around the world. You should refer to named locations that you have studied in your answer. **5**

(b) **Evaluate** the effectiveness of strategies used to combat climate change. **5**

Diagram Q9

Sea level rise required to cause city to flood at its lowest point

Venice	1 metre
Amsterdam	2 metres
San Francisco	3 metres
Edinburgh	6 metres

MARKS

Question 10 – Trade, Aid and Geopolitics

(a) Look at Diagram Q10A and Diagram Q10B.

 Account for the inequalities in trade shown.   **5**

Diagram Q10A: Employment structure

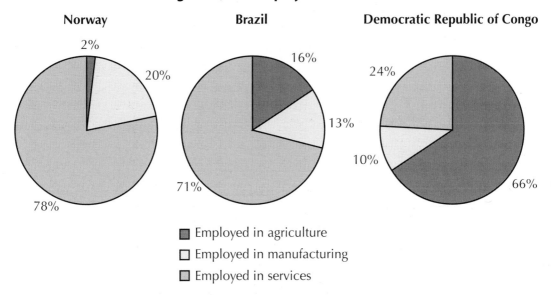

Norway

2%

20%

78%

Brazil

16%

13%

10%

71%

Democratic Republic of Congo

24%

66%

■ Employed in agriculture
□ Employed in manufacturing
■ Employed in services

Diagram Q10B: Selected development indicators

		Norway	Brazil	Democratic Republic of Congo
Trade statistics (billion US$, 2013)	**Exports**	154.2	244.8	3.1
	Imports	90.1	241.4	4.6
	Balance of trade	64.1	3.4	-1.5
	GDP per capita (US$) PPP*	55,400	12,100	600

PPP* = Purchasing Power Parity

(b) Referring to strategies you have studied, **explain** methods used to reduce trade inequalities between developed and developing countries.   **5**

Question 11 – Energy

(a) Look at Diagram Q11.

Referring to both renewable **and** non-renewable energy sources, **account for** the global distribution of energy resources. **6**

Diagram Q11: Global distribution of oil (countries proportional to oil reserves)

World reserves of oil		
	Billions of barrels	Percentage of world reserves
Saudi Arabia	262.73	22.3%
Iran	132.46	11.2%
Iraq	115.00	9.7%
Kuwait	99.00	8.4%
United Arab Emirates	97.80	8.3%
Venezuela	77.22	6.5%
Russia	72.27	6.1%
Kazakhstan	39.62	3.4%
Libya	39.12	3.3%
Nigeria	35.25	3.0%
United States	21.37	1.8%
China	17.07	1.4%
Canada	16.80	1.4%
Qatar	15.20	1.3%

(b) **Evaluate** the suitability of non-renewable approaches to meet growing energy demands across the globe. You should refer to named countries in your answer. **4**

SECTION 4: APPLICATION OF GEOGRAPHICAL SKILLS – 10 marks

Attempt the question

Question 12

Running club, Dunoon Hill Runners, have applied for permission to organise a 21-mile trail race on the outskirts of the town. As well as the full distance, the race will also have the option of a paired event with two competitors running 10.5 miles each to complete the distance.

Race Briefing – Dunoon Hill Runners 21-mile trail race

Club secretary, Kenny Taylor said:

The route has the potential to become one of Scotland's classic trail races as well as boosting the local economy. The plan is to choose a route that:

- is off road, i.e. not on A or B class roads
- provides a challenging/hilly course for competitors
- encourages business in the area around the course
- is scenic for competitors
- provides facilities for competitors at the start/finish line
- has a suitable transition area* approximately half-way along the route for those running the relay option.

*Transition area refers to the point where runner 1 will finish and pass on to runner 2 in the relay team. It should be easily accessible and allow access for vehicles."

Study Q12A: OS map; Map Q12B: Proposed start and finish area Diagram Q12A; Diagram Q12B and Graph Q12;

Referring to map evidence and the resources provided, **evaluate** the proposed route in terms of its suitability to meet the criteria set by the Dunoon Hill Runners.

You **must** also suggest a possible location near the half-way point for a race transition area. This area will be where runners who choose to participate as part of a relay team could pass on the baton. It should be easily accessible for the second runner to reach by transport and for the first runner to return to the start/finish area. **10**

Map Q12B: Proposed start and finish area, Benmore Gardens (142855)

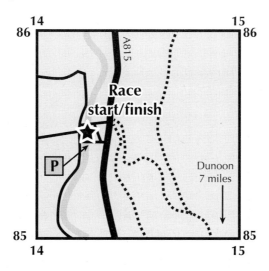

Diagram Q12A: Facilities at proposed start/finish area Benmore Gardens

- 7 miles north of Dunoon on the A815
- Café and gift shop open all year round
- Garden admission – £6 adult, £5 concession, children free
- Gardens open from the start of March through to end of October
- Highlights – giant redwood avenue (planted 1863) and formal garden
- Toilets, car parking and walking trails at the visitor centre

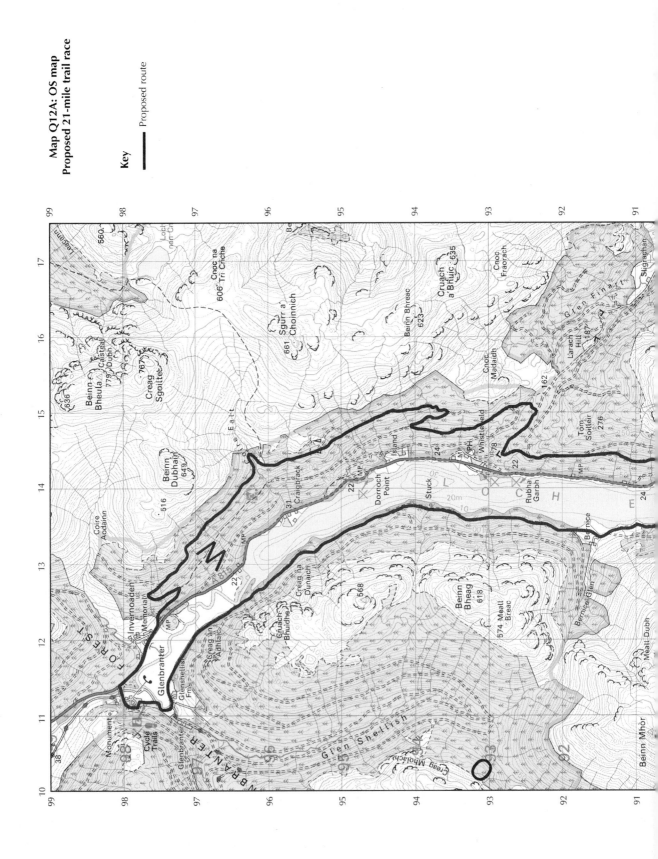

Map Q12A: OS map
Proposed 21-mile trail race

Key
—— Proposed route

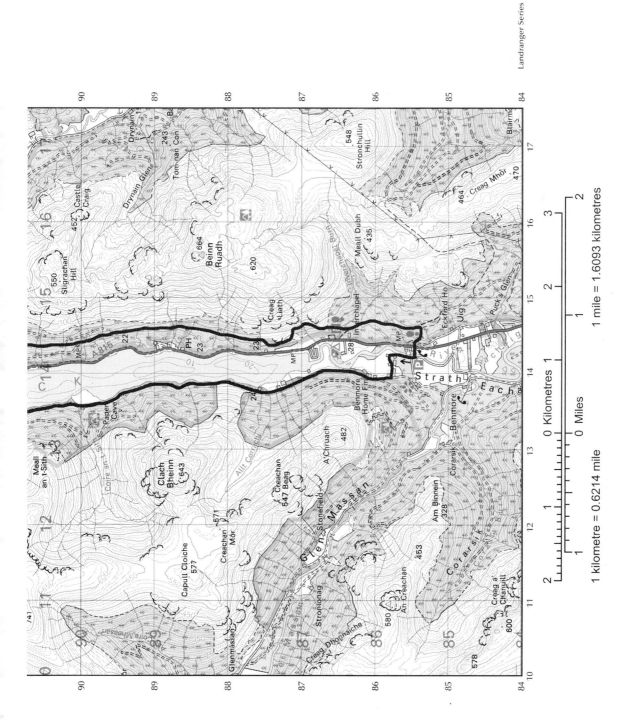

Diagram Q12B: Competitor information leaflet

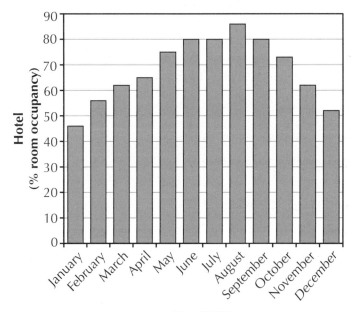

Dunoon Hill Runners
21-mile Trail Race

- 21 Mile Trail Race – Sunday 31 January
- Limited to 100 places in the first year
- Stunning Loch Side Course
- First prize male and female £100
- Medals for all finishers

For further information and race entry visit:
www.dunoonhillrunners.org.uk

Graph Q12: Hotel occupancy rate by month, Scotland 2013

Hotel (% room occupancy) by month for Year 2013:
January 46, February 56, March 62, April 65, May 75, June 80, July 80, August 86, September 80, October 73, November 62, December 52

[END OF PRACTICE QUESTION PAPER]

Practice Papers for SQA Exams

HIGHER
GEOGRAPHY
Exam C

Duration – 2 hours and 15 minutes

Total marks – 60

SECTION 1 – PHYSICAL ENVIRONMENTS – 15 marks

Attempt ALL questions.

SECTION 2 – HUMAN ENVIRONMENTS – 15 marks

Attempt ALL questions.

SECTION 3 – GLOBAL ISSUES – 20 marks

Attempt TWO questions.

SECTION 4 – APPLICATION OF GEOGRAPHICAL SKILLS – 10 marks

Attempt the question.

Credit will be given for appropriately labelled sketch maps and diagrams.

Write your answers clearly in the answer booklet provided. In the answer booklet you must clearly identify the question number you are attempting.

Use **blue** or **black** ink.

Note: The reference maps and diagrams in this paper have been printed in black ink only. No other colours have been used.

Scotland's leading educational publishers

SECTION 1: PHYSICAL ENVIRONMENTS – 15 MARKS

Attempt ALL questions

Question 1

A drainage basin is an open system with four elements.

Account for the movement of water within a drainage basin with reference to the four elements in Diagram Q1.

5

Diagram Q1: Drainage basin

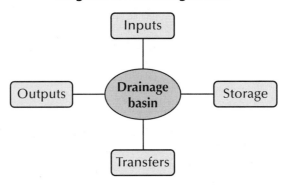

Question 2

Look at Map Q2 and Diagram Q2.

Account for the changing rainfall pattern as you move inland from Abidjan to Timbuktu.

You should refer to the position of the ITCZ within your answer.

4

Map Q2: West Africa

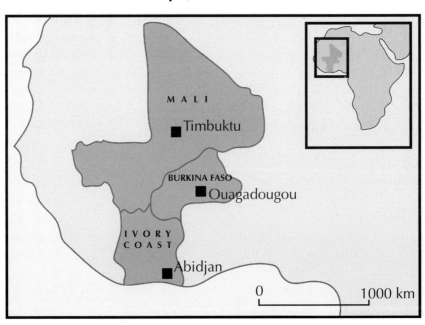

Diagram Q2: West Africa – selected rainfall graphs

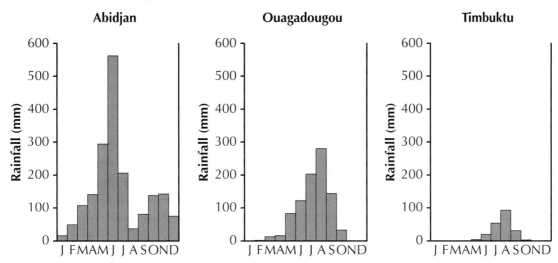

Question 3

Diagram Q3: Wind farm proposal on edge of Loch Lomond National Park

An energy company has submitted a proposal to build 20 wind turbines on a hill near Loch Lomond. The turbines would be situated on farmland and would be visible from the top of nearby Conic Hill within the National Park. Conic Hill forms part of the 96-mile long West Highland Way, a popular walking route that attracts many visitors each year.

Look at Diagram Q3.

For any named coastal **or** glaciated landscape you have studied:

    (a) **Explain** the land use conflicts that have occurred.

    (b) **Evaluate** the management strategies taken to resolve these conflicts.   **6**

SECTION 2: HUMAN ENVIRONMENTS – 15 marks

Attempt ALL questions

Question 4

Explain the techniques used to combat rural land degradation in a named rainforest **or** semi-arid area you have studied.

5

Question 5

Look at Diagrams Q5A and Q5B.

Discuss the possible consequences of the 2050 population structure for Chad.

You should refer to the impact on the economy of Chad **and** the impact on the welfare of its citizens.

5

Diagram Q5A: Population pyramid for Chad, 2010

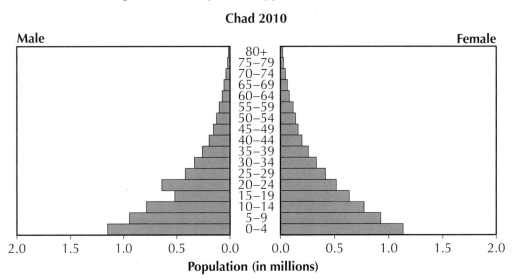

Diagram Q5B: Projected population pyramid for Chad, 2050

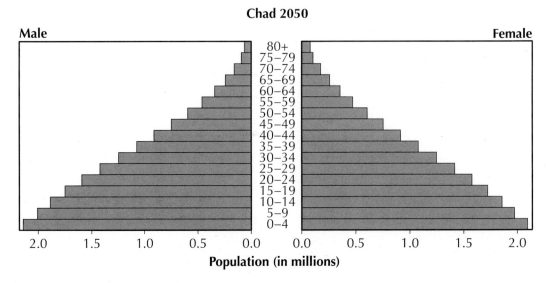

Question 6

> Cities in the developed world face many challenges at the start of the 21st century. Many of these problems are a result of changes that have occurred in the preceding 60 years.
>
> Inner city housing has changed dramatically in **developed world cities**.

With reference to a named city in the **developed world, explain** the strategies used to solve the inner city housing problem.   **5**

SECTION 3: GLOBAL ISSUES – 20 marks

Attempt TWO questions

Question 7 – River Basin Management

    (a) For a river basin you have studied, **analyse** the physical characteristics.   **4**

    (b) **Evaluate** the social, economic and environmental benefits **and** adverse consequences of a named water project you have studied.   **6**

Question 8 – Development and Health

> '*Malaria affects 109 countries worldwide… 35 countries suffer 98% of the global death toll. Just five of these – Nigeria, The Democratic Republic of Congo, Uganda, Ethiopia and Tanzania – account for 50% of global deaths and 47% of all malaria cases. Providing support to help high-burden countries … is a key priority.*'
>
> Roll Back Malaria – The Global Malaria Action Plan

Malaria, cholera and bilharzia/schistosomiasis are water-related diseases.

For **one** of these diseases **or** any other water-related disease you have studied:

(a) **Discuss** the social and economic impact this disease has on affected areas.

(b) **Explain** the management strategies used to combat the disease. **10**

Question 9 – Global Climate Change

(a) With reference to areas you have studied, **discuss** the impact global warming is having on locations throughout the world. You may refer to positive **and** negative impacts in your answer.   **5**

(b) During the last 15 years, a range of strategies have been adopted that are designed to slow down rates of climate change **and** minimise the impact of a changing climate.

Referring to strategies you have studied, **to what extent** have these strategies been effective in combating climate change? You should refer to **at least two** strategies in your answer.   **5**

Question 10 – Trade, Aid and Geopolitics

Large areas of farmland and rainforest are being cleared in Indonesia and other **developing countries.** This space is then used for cash crops like palm oil. The palm oil is exported around the world. Developments of this nature have social, economic and environmental impacts on the people and environment in the countries affected.

(a) Look at Diagram Q10A.

Many **developing countries** export raw materials such as palm oil to **developed nations.**

Discuss the impact of trade patterns of this nature on the people **and** environment of developing nations. You may refer to positive **and** negative impacts.　　**6**

Diagram Q10A: Palm oil production and rainforest cover 1964–2009

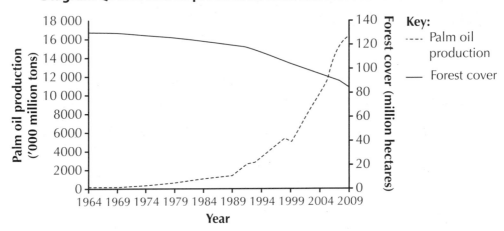

(b) Look at Diagram Q10B.

Fair trade is one method of addressing trade inequalities.

Evaluate strategies taken to reduce global trade inequalities.　　**4**

Diagram Q10B: Fair trade around the world

Producer countries

Where Fairtrade products are sold

Question 11 – Energy

Diagram Q11: World energy consumption

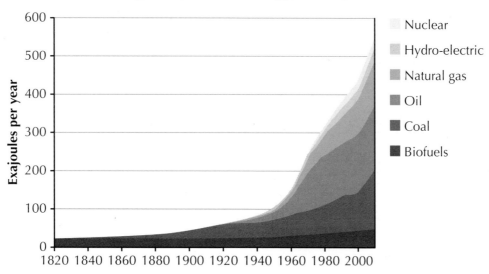

(a) **Explain** the growth in demand in energy across the world.

You should refer to factors in developing **and** developed countries in your answer.   **6**

(b) **To what extent** are renewable approaches to meeting energy needs suitable in countries you have studied?   **4**

SECTION 4: APPLICATION OF GEOGRAPHICAL SKILLS – 10 marks

Attempt the question

Question 12

Air Space is a purpose-built indoor trampoline playground on the outskirts of East Kilbride. It opened in December 2014.

Study Map Q12A: OS map of East Kilbride; Map Q12B: Central Scotland; Map Q12C: Location of Air Space; Information Q12A: Newspaper extract; Information Q12B: Facebook page and Graph Q12

Referring to map evidence and other information from the sources:

(a) **Discuss** the suitability of the **site** for Air Space UK.   **4**

(b) **Evaluate** the social, economic and environmental impact of the opening of Air Space UK.   **6**

Map Q12B: Central Scotland

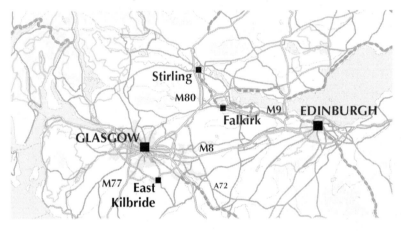

Map Q12A: OS map of East Kilbride

Landranger Series

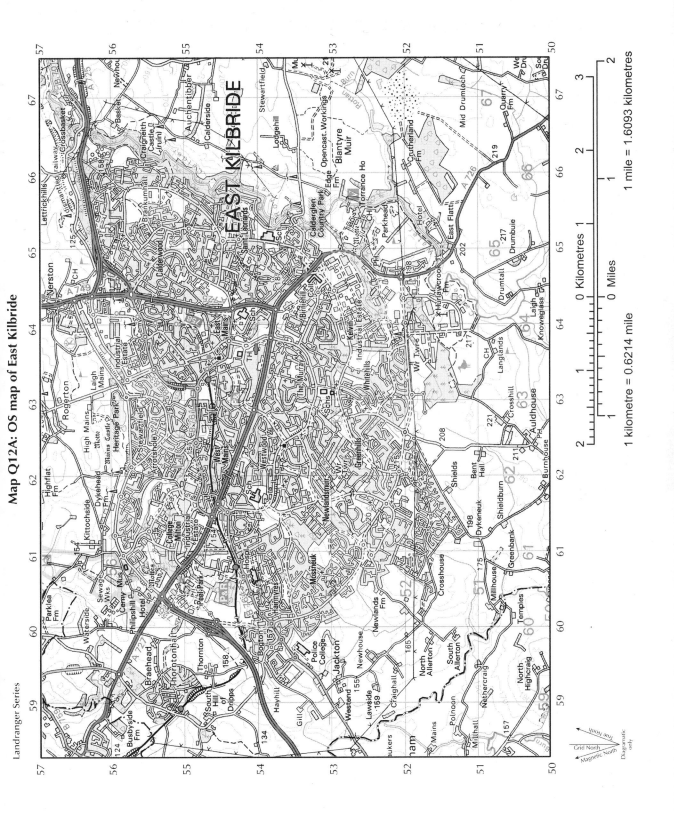

1 mile = 1.6093 kilometres

1 kilometre = 0.6214 mile

Map Q12C: Location of Air Space

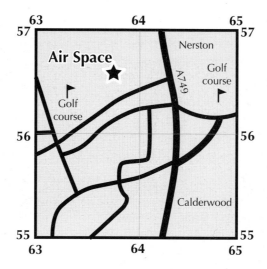

**Information Q12A: Newspaper article about Air Space UK, *Daily Record*
(26 November 2014)**

EAST KILBRIDE SPRINGS TO LIFE

A giant urban playground – with added bounce – is set to breathe new life into Playsport when it opens its doors to the public next week.

Catapulting onto the sport and leisure scene in East Kilbride, Air Space will be Scotland's first, and Europe's largest, freestyle jumping arena.

Boasting 100 interconnected, wall-to-wall trampolines, three dodgeball courts, spring-loaded football, two slamball hoops, as well as air bags, wall running and a snowsports bounce board, visitors – otherwise known as jumpsters – will be spoilt for choice when visiting the impressive £2million facility.

Information Q12B: Facebook page – Air Space

January 2015 (6 weeks after opening)

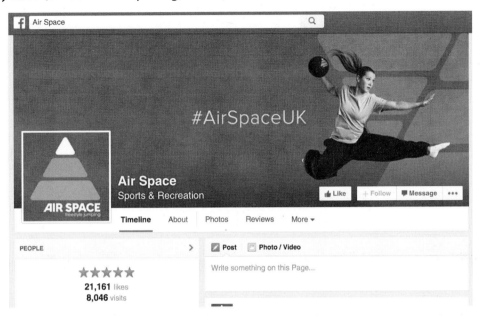

Note: People who have not visited Air Space will have liked their Facebook page. Not all visitors to Air Space UK will record this visit via Facebook.

Graph Q12: Top 10 visitor attractions in Greater Glasgow and Clyde Valley

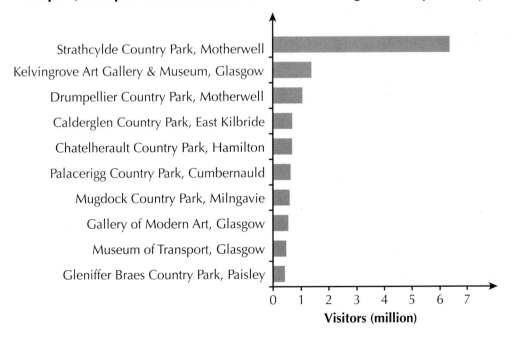

[END OF PRACTICE QUESTION PAPER]

Practice Papers for SQA Exams

HIGHER
GEOGRAPHY
Exam D

Duration – 2 hours and 15 minutes

Total marks – 60

SECTION 1 – PHYSICAL ENVIRONMENTS – 15 marks

Attempt ALL questions.

SECTION 2 – HUMAN ENVIRONMENTS – 15 marks

Attempt ALL questions.

SECTION 3 – GLOBAL ISSUES – 20 marks

Attempt TWO questions.

SECTION 4 – APPLICATION OF GEOGRAPHICAL SKILLS – 10 marks

Attempt the question.

Credit will be given for appropriately labelled sketch maps and diagrams.

Write your answers clearly in the answer booklet provided. In the answer booklet you must clearly identify the question number you are attempting.

Use **blue** or **black** ink.

Note: The reference maps and diagrams in this paper have been printed in black ink only. No other colours have been used.

Scotland's leading educational publishers

SECTION 1: PHYSICAL ENVIRONMENTS – 15 marks

Attempt ALL questions

Question 1

Explain how a gley soil is formed.

You may wish to refer to soil forming properties such as climate, drainage, natural vegetation, rock type and soil organisms.   **6**

Question 2

Explain the conditions and processes involved in the formation of **one** of the following features of glacial deposition.

- Drumlin

- Terminal moraine

- Outwash plain

You may wish to use an annotated diagram or diagrams.   **4**

Question 3

Explain why only 50% of the solar energy reaching the edge of the atmosphere reaches the surface of the Earth.   **5**

SECTION 2: HUMAN ENVIRONMENTS – 15 marks

Attempt ALL questions

Question 4

Look at Diagram Q4 and Photograph Q4

With reference to a named city you have studied in a **developing** country:

(a) **Explain** the strategies used to manage problems created by shanty towns; and

(b) **To what extent** have solutions been effective in your chosen city?   **7**

Diagram Q4: Percentage of total population living in urban areas

Region	Urban population (%)		
	1970	1994	2025
More developed regions	**67**	**75**	**84**
Europe	64	73	83
North America	74	76	85
Less developed regions	**25**	**37**	**57**
Africa	23	33	54
South and Central America	57	74	85

Photograph Q4: A shanty town in Cape Town, South Africa

Question 5

With reference to a **voluntary** migration you have studied, **explain** the causes of this migration.   **4**

Question 6

Explain the challenges faced by developing countries when collecting accurate census data. **4**

SECTION 3: GLOBAL ISSUES – 20 marks

Attempt TWO questions

Question 7 – River Basin Management

(a) For any water management project you have studied, **explain** the physical factors to be considered when selecting and developing a site for a dam.

4

(b) **Discuss** the socio-economic **and** environmental impacts of a named water management project you have studied.

6

Question 8 – Development and Health

(a) Look at Diagram Q8.

Referring to countries you have studied, **explain** why levels of development vary between countries in the **developing** world. **5**

Diagram Q8: Selected indicators of development

	Mali	Cuba
GDP per capita (US $)	739	6051
Adult literacy (%)	34	100
Life expectancy (years)	55	79
Birth rate (per 1000)	47	10

(b) **To what extent** are primary health-care measures appropriate for people in developing countries?

You should refer to named examples in your answer. **5**

MARKS

Question 9 – Global Climate Change

(a) **Account for** the human factors that may lead to the global temperature projection shown in Diagram Q9.

5

Diagram Q9: Global warming projection

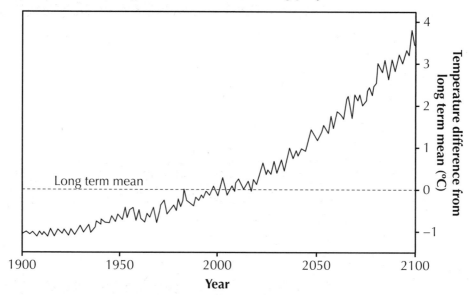

(b) **Discuss** the impact of global warming on named locations throughout the world.

5

Question 10 – Trade, Aid and Geopolitics

(a) Look at Diagram Q10.

Referring to examples you have studied, **account for** inequalities in world trade. **4**

Diagram Q10: Changing economic superpowers

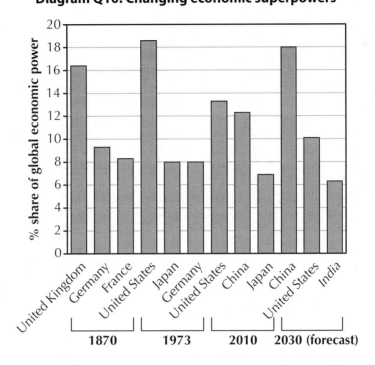

(b) With reference to named examples, **explain** the approaches used to reduce trade inequalities.

To what extent have these strategies been successful? **6**

Question 11 – Energy

(a) Look at Diagram Q11.

Explain the global distribution of energy resources with reference to renewable **and** non-renewable resources.

5

Diagram Q11: Energy sources

Coal

Oil

Natural gas

Uranium and nuclear

Hydro power

Bioenergy and waste

Wind

Solar PV

Geothermal

Peat

Marine energies

(b) Referring to named examples you have studied, **to what extent** have energy needs been suitably met by renewable approaches?

5

SECTION 4: APPLICATION OF GEOGRAPHICAL SKILLS – 10 marks

Attempt the question

Question 12

Following the Commonwealth Games in 2014, authorities in Glasgow are keen to host future sporting events to promote the legacy of the Games.

One proposal is to host an international cycling event on the route of the Commonwealth Road Race. A women's event will start at 9 am and will be a 100-km, 7-lap race of the 14.4 km circuit. A men's race will start at 2 pm and will be a 172-km, 12-lap race of the 14.4 km circuit.

On the day before the race, an open event will allow members of the public to try the course. Members of the public will be allowed to cycle on the course from 4 pm to 7 pm.

The brief is outlined below.

Brief for Glasgow Cycle-Thon

The route should:

- be spectator friendly and accessible at various points for those who want to come out and watch the race

- pass important tourist landmarks to promote the city to television audiences

- be suitable for all participants who may want to try the course on the day before the race

- start and finish at a suitable location for competitors

- minimise the need for road closures.

Study Map Q12A: OS map of Glasgow; Map Q12B: Proposed route Glasgow Cycle-Thon; Map Q12C: Map extract; Diagram Q12; and Transect Q12: Route profile.

Read the brief for the proposed Glasgow Cycle-Thon. **Evaluate** the suitability of the proposed route for an event of this nature (Map Q12B). In your answer you should **discuss** how the organisers could improve the route.

You must refer to map evidence in your answer.   **10**

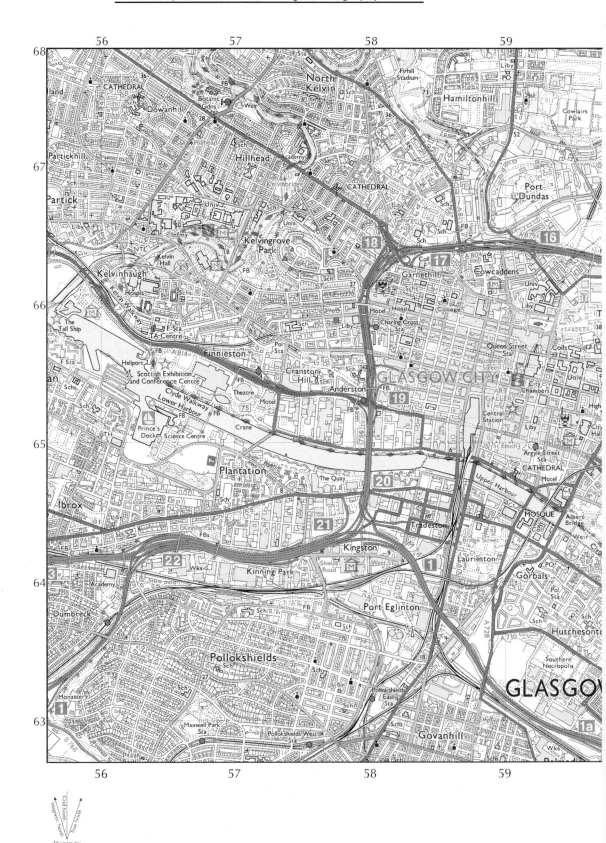

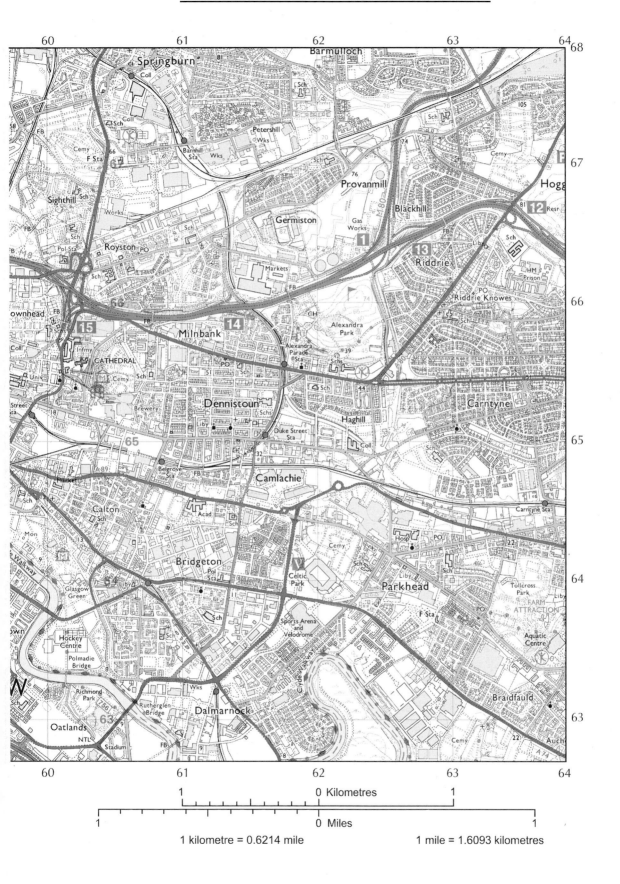

Map Q12B: Proposed route Glasgow Cycle-Thon

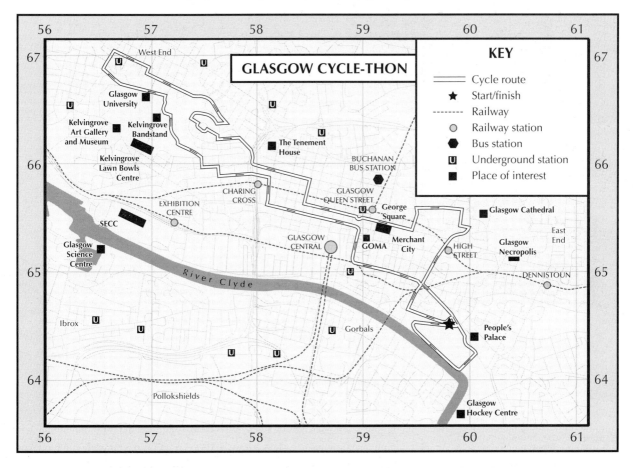

Map Q12C: Map extract

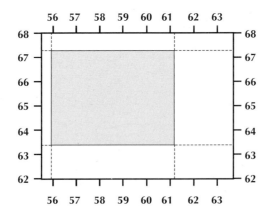

MARKS

Diagram Q12

GLASGOW CYCLE-THON

LAST WEEKEND IN JULY

See international cycle stars

Try the Commonwealth Course on
Saturday – families welcome

Safe route along closed roads

Glasgow 2014 Commonwealth Champion Geraint Thomas

Transect Q12: Route profile

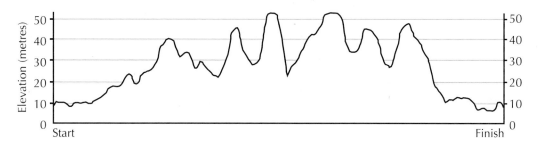

[END OF PRACTICE QUESTION PAPER]

General marking principles

The following guidelines are issued to markers. The language has been modified to make them student friendly and give you an idea of how the marker will approach your script. Bear these points in mind as you correct your work.

(a) Marking will always be positive – marks are awarded for demonstrating relevant skills, knowledge and understanding. No marks are deducted for omissions or mistakes.

> ### TOP TIP
> During the exam if you want to change your answer do not waste time scoring out what you have already written. Rather, leave two lines and make your next point. Scored-out work is not marked, so if you leave it there is a chance you might get some credit. You will not lose any marks if it is wrong.

(b) Where candidates answer two parts of a question when the instruction is to answer one part only, markers should mark both parts and award the higher mark.

> ### TOP TIP
> This is a trap you do not want to fall into as the time spent answering more than one part is unlikely to be made up when half of your answer is discounted. Look very carefully for command words such as:
>
> - **and** – indicates that you should answer both parts
>
> - **or** – you should only answer one part
>
> - **either** – choose from the list, i.e. do not do all.

(c) The marking scheme is not a comprehensive list. You can get marks for points you make that do not appear in the marking scheme as long as they are fully developed and relevant. The answers on the following pages are only *possible* answers.

(d) Points must relate to the question asked. Where points *do not* relate to the context of the question, these will not be credited.

> ### TOP TIP
> This is a common error for students to make due to exam pressure. Rushing to finish also means it is very easy to misread questions and lots of good points are unrewarded as they do not relate to the question. Be very careful. Read questions twice, underline key words and consider making a 5–10 word plan before writing your answer to keep your answer focused.

(e) To achieve a mark in Higher Geography you need to:

- make a point that is relevant to the question

- develop this point by giving sufficient detail, examples, reasons and/or evidence

- respond to the command word within the question – e.g. are you being asked to evaluate, explain or account for?

TOP TIP

Top Tip boxes appear after some of the answers in the marking scheme. They contain advice on how to approach different types of question. Marking instructions in this book will also add to your revision notes in some topics.

Take your time when you reflect on your own work. Plan well ahead and build in time to try questions again as you approach important deadlines.

Although there are more tips for the first two question papers, the tips apply to a number of questions and are not unique to the questions they appear next to. Following these tips will refine your exam technique and help you perform to your potential in the final exam.

Mark scheme for Exam A

In this section of the book you will find additional instructions for marking questions. For example, in some questions you may be told that a maximum of 4 marks can be awarded for one part of the question. Pay close attention as these are guidelines for markers and instructions like this will apply in the final exam.

SECTION 1: Physical Environments

1. Possible answers might include:

   - Headlands and bays are found in areas where bands of hard and soft rock lie perpendicular (discordant) to the coastline. (1 mark)

   - As the result of the process of differential erosion, the softer rock is eroded more easily and at a faster rate than the hard rock. (1 mark)

   - Hydraulic action may occur when the crashing water of the waves forces air into small cracks in the rock causing mini-explosions. The repeat action of these explosions causes rock to fracture and break away. (1 mark)

   - Abrasion, when the load carried by the waves is thrown against the coastline, wears away at the coastline, breaking off material. (1 mark)

   - At the location of the soft rock, material is removed in transportation by the waves, leaving an area of the coastline set back from the surrounding harder/more resistant rock. This forms a bay. (1 mark)

   - Bands of hard rock are more resistant and remain as headlands at one or both ends of a bay. (1 mark)

   > ### TOP TIP
   > You should include annotated (labelled) diagrams when explaining the formation of physical features. It is good practice to include a before, during and after diagram for each feature. Keep diagrams simple – they do not need to be 3D.
   > Including named examples demonstrates good wider knowledge and takes very little time. For example, Swanage Bay is enclosed by headlands at Ballard and Peveril Point. You could include these names on your diagram.

2. Possible answers might include the following:

   - The movement of the ITCZ is connected to the arrival of seasonal rains in West Africa. (1 mark)

   - A surplus of energy leads to the warming of air near the equator. This air becomes less dense and rises. (1 mark)

   - At the ITCZ, warm air rises and condenses, forming tall clouds with associated rainfall. Maximum monthly rainfall is associated with the ITCZ being overhead. (1 mark)

- The ITCZ moves north during the summer months bringing rain with it. As a result, places to the south of the ITCZ come under the influence of the Tropical Maritime air mass. This brings hot and humid conditions for longer periods of the year. (1 mark)

- In the southern part of West Africa, total annual rainfall is higher due to the presence of this air mass. (1 mark)

- Two peaks of rainfall are common nearer the coast of West Africa due to the movement north of the ITCZ in the summer and the return of the ITCZ en route to the equator in the winter. (1 mark)

- In the northern part of the region, total annual rainfall is significantly lower. Here a long dry season dominates because of the tropical continental air mass, which brings hot and dry conditions. (1 mark)

> **TOP TIP**
> This question asks you to consider the impact on rainfall in West Africa. This is a compulsory case study so you must know it well.

3. Answers should refer to:

- the shorter lag time in urban areas vs. the longer lag time in rural areas

- the steeper rising limb in urban areas vs. the shallower rising limb in rural areas

- the higher peak discharge in urban areas vs. the lower peak discharge in rural areas

- the steeper falling/recession limb in urban areas vs the shallower falling limb in rural areas

- the shorter time to return to base flow in urban areas vs the longer time to return to base flow in rural areas.

> **TOP TIP**
> Note that this question type requires you to **account for**. This means that you have to give reasons. The information above is descriptive. Answers that rely on description will not score highly and will not achieve over half marks. Use this question to spot the difference between points that are descriptive (above) and those that account for/explain (see below).

Possible answers might include:

- The **lag time** (duration between **peak rainfall** and **peak discharge**) is shorter in the urban hydrograph as **impermeable** surfaces (e.g. tarmac) lead to rapid **surface run-off** via drains and sewers resulting in quick **transfer** of water to the river. (1 mark)

> **TOP TIP**
> Using the correct **terminology** improves the sophistication of your answer. For example, use terms like rising limb, recession limb, surface run-off and interception in this answer. Appropriate terminology is highlighted in bold above.

- Water management systems in urban areas move water quickly into the river and sewers leading to a steep rising limb. (1 mark)

- In rural areas interception by vegetation acts as a store for rainfall, delaying the transfer of water to the river and leading to a longer lag time. (1 mark)

- Infiltration is greater in rural areas, resulting in transfer via groundwater flow. This process slows the journey of rainfall into the river, resulting in a shallower recession limb. (1 mark)

- Percolation and groundwater storage are common in rural areas with permeable surfaces. This leads to a slow release of water in the hours, and even days, after a rain storm which delays the return of rivers in rural areas to normal levels (base flow). (1 mark)

- The slow transfer of water results in a lower peak discharge in rural areas. (1 mark)

SECTION 2: Human Environments

4. Possible answers might include:

   - Scotland is facing issues associated with an ageing population.

   - As people live longer, countries face issues with an increasingly elderly population in terms of providing them with pensions, as the cost of this service increases. (1 mark)

   - In many countries, governments are looking at increasing the retirement age, in order to reduce the number of pensioners and pension contributions they have to make. (1 mark)

   - The elderly use some services disproportionately – health-care costs will increase dramatically due to the demands of an ageing population. (1 mark)

   - Social services – including care homes, meals on wheels and free public transport – will have to be provided to larger numbers of pensioners, again increasing the financial cost. (1 mark)

   - These developments may lead to greater employment prospects in these sectors and result in economic growth. (1 mark)

   - A growing elderly population can contribute effectively via informal childcare arrangements for grandchildren, allowing more working-age adults to pursue employment. (1 mark)

   - Falling birth rates lead to fewer children. This may result in the closure of services such as pre- and post-natal care, pre-school and education, resulting in a loss of jobs in these sectors. (1 mark)

   - In the future, there will be fewer workers to support a growing elderly population. A higher dependency ratio will make it difficult to raise the tax revenue required to meet the needs of the elderly dependents. (1 mark)

   - Governments have decided or may decide to encourage further migration of working-age adults to balance their dependency ratio. This may lead to tension and social issues. (1 mark)

> **TOP TIP**
> Case studies are important. The best candidates are able to recall information in the form of statistics, dates etc. in their answers. Before the exam, research two or three key facts for each of your case studies and use them if appropriate. Share these with others in your class to build up a good selection of facts.
> Remember that the marker has taught Higher Geography and has likely covered the same case studies as you – do not be tempted to make up information, it is unlikely to be accurate and will be checked. Make sure information comes from reliable sources.

5. For full marks you must comment on the effectiveness of the techniques. Answers must refer to named examples from a developed world city. Candidates who do not will not score full marks. Answers will vary depending on the city studied. For example, for Glasgow, possible answers might include:

   - The recently opened extension to the M74 has removed a large volume of traffic, easing congestion on the Kingston Bridge. (1 mark)

   - Measures such as higher on-street parking charges in the CBD discourage people from bringing their cars into the centre of the city, limiting on-street parking/lane blockage. (1 mark)

   - This is targeted at normal office hours when congestion is worst and reverts to free parking from 6 pm to 8 am when congestion is lower. (1 mark)

   - Shops receive most of their deliveries early in the morning – when lorries and trucks are allowed on pedestrianised streets, e.g. Buchanan Street – removing commercial vehicles from the centre of the city. (1 mark)

   - Public transport has been improved to encourage people in Glasgow to leave cars at home. For example, bus frequency on major routes has been improved to every 10 minutes at peak times. (1 mark)

   - More carriages have been added to trains and wi-fi added to make the option of public transport more appealing to commuters and reduce the number of cars travelling into the CBD. (1 mark)

   - Bus lanes (e.g. on Maryhill Road) reduce journey times by public transport and make it more attractive than travelling by car. (1 mark)

   - This has been partially successful but can lead to heavier congestion on major roads at rush hour as there is less road space for cars, which remain the most common vehicle. (1 mark)

   - Season tickets and zone cards offer discounts for those using public transport to commute regularly, but campaigners argue that these remain too expensive to be truly effective in attracting people to leave their cars at home. (1 mark)

   - Park and ride schemes, e.g. at Shields Road, allow people to travel part of the way by car but to leave their car outside the CBD and complete their journey using the underground/ overground train network. (1 mark)

- The media is used to alert the public of potential problems to prevent congestion from building at hotspots, e.g. accidents or road works. For example, travel news on radio stations, using electronic noticeboards on main roads/motorways and using social media such as Transport Scotland updates on Twitter. (1 mark)

- However, these methods can lead to the build-up of traffic in other areas as large numbers of commuters seek to avoid the original incident. (1 mark)

> ### TOP TIP
> Candidates find it difficult to comment on the effectiveness of solutions or strategies. There are a number of good examples above. You need to say whether something has dealt with the problem or not and give reasons. There are usually points of view on both sides of the issue and it is worth noting these.
> It is not enough to say this solution has been very effective. In summary, give an overall impression. In this context, cities like Glasgow still have significant congestion problems, largely at rush hour, as large numbers of commuters try to access the CBD at the same time of day during the week. The lack of bridging points exacerbates this issue in Glasgow and the Kingston Bridge remains a major bottleneck.

SECTION 3: Global Issues

6. (a) Answers will depend on the case study chosen. For example, for the Colorado River possible answers might include:

   - The climate in the basin is characteristic of a hot desert with low rates of precipitation below 250 mm per annum. The rain that falls is unpredictable resulting in seasonal flow in the river. (1 mark)

   - Melt water from snow at higher altitude in the upper course of the basin (Wyoming and Colorado) feeds the river. Snowmelt provides the majority of water in the basin. (1 mark)

   - When precipitation does fall thunderstorms are common, resulting in a risk of flooding caused by a rapid increase and change in the amount of water within the basin. (1 mark)

   - Poor soils are common, which limits the breadth of vegetation found in the basin. This reduces interception and results in rapid transfer of rainwater into the river. (1 mark)

   - Temperatures can rise above 40°C, leading to high evaporation rates, particularly in the summer. Water loss from vegetation is also high but species have adapted to minimise water loss. (1 mark)

   - The geology of the area and climatic conditions lead to landscapes characterised by onion-skin weathering as a result of thermal expansion and contraction of the predominant sandstone rock. (1 mark)

> ### TOP TIP
> A good strategy in a question like this is to consider different factors to discuss. If you can remember these it will ensure your answer has good variety. For example, in this question type the examiner will be looking for ideas about climate, geology, soil/vegetation and river flow/discharge.

(b) Candidates must explain the need for water management using the prompts provided. No marks will be awarded for purely descriptive points lifted from the sources. Possible answers might include:

- There is very low rainfall in Egypt all year round; it is an arid climate, meaning management strategies are required to ensure people have drinking water throughout the year. (1 mark)

- Higher rainfall in the upper course of the river basin indicates that management of the basin is required to prevent flooding in populated areas along the river. (1 mark)

- Egypt's population is projected to continue to grow at a rapid rate, adding to the demand for water, so management is required to ensure there are adequate supplies to meet this need. (1 mark)

- Water management provides water for farming. Irrigation allows crops to be grown in the warm, dry climate. This would not be possible without water management. (1 mark)

- The Aswan Dam may provide hydroelectric power to meet the needs of the growing population and associated industries. (1 mark)

7. (a) Answers must explain why primary health-care strategies are appropriate in developing countries. Possible answers might include:

- Barefoot doctors – locals with experience in treatment of common ailments – are effective as they can reach people in remote areas far from the limited health-care that is often found in urban areas. (1 mark)

- Vaccination programmes (e.g. polio, measles) aimed at children are effective as these measures are preventative and minimise the need for costly treatment later on should someone contract one of these diseases. (1 mark)

- An additional benefit is that it can minimise disruption to education and the vital role that children play in supporting the family through their domestic roles. (1 mark)

- Cheap solutions are available for common ailments, allowing these medicines to be provided in poorer developing countries. For example, the use of oral rehydration therapy prevents complications/fatalities from diarrhoea. (1 mark)

- Education is very important and plays/songs and information displayed in pictures have been effective in overcoming difficulties faced by those who are unable to read/write. (1 mark)

- Travelling clinics/specialists are often used who can visit different areas and treat people who are unable to travel for financial and/or health reasons. (1 mark)

- Local workers are often involved by charities in the construction of local clean water projects/toilet facilities. Materials are provided and training/transferable skills encourage wider usage in the region. (1 mark)

> **TOP TIP**
> Avoid simply describing strategies. Your comments on effectiveness must be well-developed and varied. For example, 'it is cheaper' is not well-developed and should not be used more than once in your answer.

(b) Candidates must refer to both human **and** environmental/physical factors to achieve full marks. Answer will vary depending on the disease chosen, but for malaria possible answers might include:

- The female anopheles mosquito carries the disease and must be present. Its natural habitat is in areas with humid climates/temperatures between 15°C and 40°C. (1 mark)

- Areas of vegetation are required. Mosquitoes use these as resting sites where they digest. (1 mark)

- Anopheles mosquitoes breed close to stagnant water as they require this type of environment to lay their larvae, e.g. ponds. (1 mark)

- Construction by humans, e.g. reservoirs and other man-made stores, irrigation channels and basic sewage works, often increases the range of breeding grounds as an unintended consequence of increasing the surface area of stagnant water. (1 mark)

- Malaria spreads in areas of settlement where humans act as a blood reservoir. This allows mosquitoes to feed and reproduce successfully. (1 mark) The spread of malaria is increased where people work outside with exposed skin at dusk and dawn when mosquitoes are most active. (1 mark)

- The increase of tourism to afflicted areas can lead to the disease moving into new territories, leading to further spread of malaria. (1 mark)

> **TOP TIP**
> Remember subheadings can be useful to structure your work and ensure you answer **both** parts of the question. In some instances however, factors can be both human and physical, e.g. stagnant water in the question above. Marks are available for different ideas about stagnant water:
> - As a human factor, stagnant water often results from the installation of irrigation systems or drainage channels. Uncovered domestic water barrels also provide stagnant water for breeding.
> - As an environmental/physical factor, stagnant water may be found in areas where swamps or ponds are present.

8. (a) Only answers that refer to **physical** factors should be credited. Possible answers might include:

- Dust and ash released after major volcanic eruptions can block incoming solar radiation, leading to a global cooling. (1 mark)

- Increase in sunspot activity leads to above-average solar radiation reaching Earth, leading to periods of above-average temperatures. (1 mark)

- Changes in the Earth's orbit, linked to Milankovitch cycles, lead to variations in the Earth's distance from the sun and the tilt of its axis. These factors occur on regular cycles and are linked to periods of change in average temperatures. (1 mark)

- Diminishing ice cap coverage lowers global albedo rates and loss of solar radiation, increasing global temperatures. (1 mark)

- In Tundra regions, rotting vegetation and/or a thawing of the permafrost has led to the release of methane into the atmosphere. Methane is a powerful greenhouse gas and traps solar radiation in the atmosphere, warming the planet. (1 mark)

(b) Measures taken to combat climate change fall into two groups. Those that aim to reduce emissions and those that seek to minimise the impact of climate related changes. Possible answers might include:

- The Scottish Government have set ambitious targets to meet energy needs from renewable sources that do not produce greenhouse gas emissions. Scotland has focused on wind, wave and tidal approaches. (1 mark)

- International proposals, including the Kyoto Protocol, are designed to ensure that all countries comply as climate change is a cross-border/global threat that requires international cooperation. (1 mark)

- Education seeks to promote personal responsibility. Information/campaigns have focused on important issues such as food miles and recycling to reduce the individual carbon footprint. (1 mark)

- People are encouraged to lower their energy demands through a variety of ways, e.g. walking or using public transport, flying less frequently for business, insulating our homes and switching off lights and appliances when not in use. (1 mark)

- Some governments have increased taxes in certain areas to incentivise energy/ resource savings, e.g. carbon levies on flights. (1 mark)

- Many private organisations and charities are engaging in afforestation schemes to rebuild forests as carbon stores (trees breathe in carbon dioxide and breathe out oxygen). (1 mark)

- Approaches to minimise the impact of climate change include flood prevention such as the Thames Flood Barrier, which minimise the impact of climate-change-related weather events. (1 mark)

- The growth of air travel and car ownership around the world (particularly car travel in developing countries) has seen a massive increase in emissions, making it difficult to cut emission levels. (1 mark)

- Suitable options that run off renewable sources or are more energy-efficient have been relatively unpopular with customers, although there has been some success with electric cars. (1 mark)

> **TOP TIP**
> Supporting evidence from other topics might enhance the quality of answer in a question of this nature. In the Global Issues topics, a very good answer may briefly refer to these links as well as bringing in relevant and up-to-date examples from their own research.

9. (a) Possible answers might include:

- Developed countries have traditionally imported raw materials and turned these into manufactured goods for export (e.g. cars). These products are more expensive, resulting in a larger share of trade by financial value. (1 mark)

- Raw materials are exported by developing countries. They face competition from other developing nations who are keen to earn foreign currency. This process keeps prices low and limits their share of global trade. (1 mark)

- Current patterns of trade and historical trading patterns have led to concentrations of wealth in the developed world. Therefore a high proportion of trade is between developed countries further increasing their share. (1 mark)

- Trade is linked to the performance of the global economy and the value of trade rises and falls on the basis of this. During the recent economic crisis, global trade fell dramatically. (1 mark)

- China's share of world trade continues to grow faster than any other region of the world. China has been a low-cost base for manufacturing because of cheap labour costs, but is now moving into a service-based economy with a growing market for raw and processed goods. (1 mark)

- India and Africa as trade blocs have also seen their share of world trade rise. This is linked to a rise in population, affluence and demand in developing nations. (1 mark)

(b) Impacts may be positive and/or negative. Candidates could refer to named examples in their answers. Possible answers might include:

- International trade has seen investment by Trans-National Corporations (TNCs) in developing countries. Locals have benefitted from employment opportunities. Some believe this provides a more secure income than farming. (1 mark)

- Trade allows countries to earn foreign income which can be used to pay off debt. (1 mark)

- TNCs may improve local infrastructure, including electricity, sanitation, roads and communications technology, which benefits local communities. (1 mark)

- Jobs are often poorly paid and do not give enough money for workers to pay for education for their children. (1 mark)

- Many children are involved in primary or low-technology manufacturing jobs, limiting their prospects and the chances of development in the future. (1 mark)

- Working conditions are often poor with reports of little or no holidays, long hours, workers not seeing their families and dangerous buildings. (1 mark)

- For example, a factory collapsed in Bangladesh in 2013 killing more than 1000 workers. (1 mark)

- Cash crops are often grown to earn foreign income. Many of these crops, e.g. palm oil, are grown on the most fertile land or forest is cleared for agricultural land. This lowers food yields and damages habitat for wildlife. (1 mark)

- In rainforest areas, indigenous ways of life are being threatened due to contact between developers and tribespeople. (1 mark)

- Profits from ventures by TNCs often flow abroad to the home economy of the company, leading to criticism that companies are exploiting cheap labour in developing countries. This limits economic growth in developing nations. (1 mark)

> **TOP TIP**
> Giving both positive and negative impacts increases your options when answering a question of this nature.

10. (a) Possible answers might include:

- The total population in developing countries is growing rapidly. Economic growth and developments in infrastructure mean that the demand for energy is rising quickly. (1 mark)

- The demand for energy comes from both domestic and commercial users. For example, home owners seek energy for lighting, heating and electrical appliances as they become more affordable and people look to improve their own living standards. (1 mark)

- Manufacturing is growing rapidly in many countries. As in the industrial revolution in developed nations, these factories require a lot of energy to produce their goods. (1 mark)

- These products are increasingly exported, creating a large carbon footprint due to transport by air and sea freight. (1 mark)

- Developments in infrastructure and growing levels of wealth create a market for transport including the growth in car ownership, e.g. in China, thus causing more energy use. (1 mark)

(b) Candidates must refer to the effectiveness of each non-renewable approach mentioned. 1 mark can be awarded where candidates refer to **two** specific named relevant examples. Possible answers might include:

- **Coal** has been effective in countries where plentiful supplies are found and are easy to extract. Due to the geology in the UK, coal was abundant and was used to power the industrial revolution. (1 mark)

- For example, coal was easily mined on the floors of the Welsh valleys, e.g. Rhonda Valley. (1 mark)

- Concerns about coal centre around the release of carbon dioxide when it is burned for energy. This contributes to the greenhouse effect. (1 mark)

- Today coal is more commonly imported because British mines began to close in the 1970s/80s. There is a call to reopen a number of mines to meet energy demands as the infrastructure remains and this may be a cheaper option. (1 mark)

- New discoveries of coal deposits mean that coal is an option to meet energy needs. (1 mark)

- **Oil** is a major source of energy in oil-rich countries like the USA, where large deposits are found in states such as Alaska and Texas. (1 mark)

- The technology to convert oil into energy is now embedded in everyday life and appliances such as the car make it cost-effective to use this energy source. (1 mark)

- As a finite resource oil will run out and it is not a long-term solution to meeting the world's energy needs. (1 mark)

- Concerns have been raised about ongoing exploration for new sources of oil in some of the world's most environmentally sensitive areas, including the Arctic. (1 mark)

- **Natural gas** meets a large share of our energy needs. In Europe much of this gas in transported by pipelines from Eastern European countries, including Ukraine. Concerns about political stability in this region mean it is not guaranteed to meet energy needs in the area. (1 mark)

- When countries are importing energy they are vulnerable to price fluctuations, which make planning difficult. (1 mark)

- As a result, fracking in some areas is being explored as an alternative source to meet energy needs, for example in the area around Falkirk and Grangemouth. (1 mark)

> ## TOP TIP
> Across the course there is great scope to refer to current affairs. Watch the news and link in to current local, national and international stories. Do not give huge amounts of detail but show the examiner that you have a good understanding of Geography from wider reading.

SECTION 4: Application of Geographical Skills

11. (a) Possible answers might include:

- Visitors are attracted by the stunning coastal scenery such as Lulworth Cove (827798), beaches and dramatic cliffs. (1 mark)

- Ancient monuments such as the chapel and tumuli(s) offer a different type of visitor attraction alongside the Heritage Centre, which attracts people interested in the history of the area. (1 mark)

- There is a coastal path marked on the map and walkers are likely to use West Lulworth as a stopping point and may use local services such as shops and the pub. (1 mark)

- There is a camp site and caravan park (810806), which provides somewhere for visitors to stay in the area. (1 mark)

TOP TIP

Most maps used in the exam will be at the 1:50,000 scale but it is important that you are comfortable working with 1:25,000 also. Take your time with grid references.

Remember to use all of the sources on the exam paper and the map (this means you should be using grid references).

Grid references are not the only map evidence you can include. You could mention road names, names of physical features or comment on the relief of the land.

(b) Possible answers will depend on the site chosen but for Site A might include:

- The relief of the site is flatter, making construction easier and pitching caravans and tents will be more comfortable on the flatter surface. (1 mark)

- The views of the coastline at Site A are uninterrupted but at Site B views are obstructed by the relief of the land and forestry cover. (1 mark)

- Site B is much closer to the other camp/caravan site, and although a new site is required this might lead to confusion and competition between different businesses affecting profits at the existing site. (1 mark)

- Site A is closer to the coastal path and will be easier for walkers who are looking to camp to access the path. Passing walkers during the day may also stop for supplies/ lunch in the café, generating additional revenue at Site A. (1 mark)

- The site is also closer to services in the local village, such as the public house and church, providing easier and quicker access for visitors who may wish to visit these services. (1 mark)

- Farming is an important employer in the area and Site B lies in the vicinity of a farm. This may lead to problems for the farmer if visitors damage walls or leave gates open, allowing animals to escape. (1 mark)

- Visitors with dogs may cause alarm to farm animals and might lower productivity and profits on the farm. (1 mark)

- Alternatively, the farmer may be able to offer produce to visitors and increase their revenue, particularly during the busy summer months. (1 mark)

Candidates may also refer to negative factors about their chosen site. Possible negative points about Site A might include:

- Traffic will pass along narrow roads at both locations but at Site B the road is more permanent. At Site A, traffic and caravans will have to pass by buildings (which may be houses) and the noise may disturb local residents. (1 mark)

- Site A is next to the military training area. The military is an important land user in the area and this may lead to conflict as visitors may complain about the noise from the training exercises. (1 mark)

> ### TOP TIP
> In a choice-based question you should speak about all of the proposed sites. Include positive and negative factors (you might mention negative factors even for the site you ultimately choose).
> Structure is very important. Deal with it factor by factor rather than site by site. For example, it is better to write about the relief of both sites in one sentence and then the transport access in the next rather than writing one paragraph about Site A then one about Site B. Comparative language (better, flatter, easier) is useful.

Mark scheme for Exam B

SECTION 1: Physical Environments

1. Candidates must mention both the atmosphere **and** oceans to score full marks. A maximum of 4 marks should be awarded to either. Possible answers might include:

    *   Oceans and the atmosphere move water from areas of surplus at the equator towards areas of deficit at the poles. (1 mark)

    *   Cold currents, e.g. the Labrador Current, are more dense and sink while warm currents rise. This creates a conveyor belt effect moving energy towards the poles. (1 mark)

    *   Ocean currents in the Atlantic are moved in part by prevailing winds and the deflection off significant landmasses. For example, the Equatorial Current is moved north and south from the equator after colliding with South America. (1 mark)

    *   Atmospheric cells (the Polar, Hadley and Ferrel Cells) circulate air in the atmosphere. Warm air is less dense and rises at the equator before cooling and spreading 30° north and south, where it sinks. (1 mark)

    *   Some of this warm air travels towards the poles as surface winds. This occurs as air moves from areas of high pressure (30° north and south) towards areas of low pressure (60°) therefore transferring energy from the equator towards the poles. (1 mark)

    *   The remainder returns to the equator to be reheated, thus completing the Hadley cell (thermally direct). A zone of divergence occurs at this point. (1 mark)

    *   At 60°, warmer air travelling towards the equator meets cold air travelling across the surface of the Earth. At this point, the two air masses converge and rise up into the atmosphere. (1 mark)

    *   Surplus energy is carried in the warmer air towards the poles where it sinks upon cooling, creating the Polar Cell. This cell is thermally direct. (1 mark)

    > **TOP TIP**
    > You do not always have to be instructed to draw a diagram to benefit from including one. In this question, a diagram showing the three-cell model and associated air masses will make it easier to explain and enhance your answer.

2. Possible answers might include:

    *   Podzols have slow breakdown of their litter, leading to formation of an acid mor humus while brown earth soils have faster breakdown, creating a mildly acidic mull humus. This is due to the higher temperatures in regions where brown earth soils are found, aiding the breakdown of material. (1 mark)

    *   Furthermore, coniferous vegetation is waxy and acidic which hinders decomposition. (1 mark)

- The podzol soil has clearly defined horizons whereas the brown earth has mixed horizons that are indistinct. Greater levels of organic matter in the brown earth are responsible for this mixing. (1 mark)

- Brown earth soils have a brown colour while podzols are grey and have a red/dark brown iron pan. This is caused by higher rates of leaching in podzols. (1 mark)

- In podzols this is linked to melt water, which creates excessive spring leaching. (1 mark)

- Podzols can become waterlogged as the iron pan may impede drainage whereas deep root penetration allows water to pass more freely through brown earth soils into the parent material. (1 mark)

> **TOP TIP**
> In comparison questions you should consider structure carefully. You could write a paragraph about each soil type. However, marks are awarded for direct comparisons so you must talk about the horizons and organic matter etc. in both soils; i.e. don't just talk about the horizon in one soil as you will not get marks. Therefore, a better structure to your answer would be property-by-property rather than soil-by-soil.

3. Responses will depend on the area studied. Possible answers for the Dorset coast might include:

   - Tourism at honeypot sites such as West Lulworth can create congestion, particularly during public holidays and summer months. Additional traffic on narrow country roads inconveniences local residents and businesses. (1 mark)

   - Footpath erosion is common due to the volume of visitors to the Jurassic Coastline. Erosion on the South-West Coastal Path creates unsightly scars on the landscape. (1 mark)

   - Tourism often comes into conflict with farming as visitors lead to an increase in the frequency of open gates and dogs in the area. Both can harm farm animals. (1 mark)

   - Wild animals may choke on litter dropped by visitors, including discarded plastic food containers which attract animals. (1 mark)

   - Hotels, visitor centres and purpose-built services may not fit in with the natural landscape, leading to visual pollution via the built environment. (1 mark)

SECTION 2: Human Environments

4. Problems are likely to focus on issues surrounding the growth of and conditions within shanty towns. Candidates may answer part (a) and (b) together. Answers that do not evaluate impacts should score no more than 4 marks. Answers will vary depending on the city studied, but for Mumbai possible answers might include:

   - In cities such as Mumbai, new residents arrive daily into slums such as Dharavi from the surrounding countryside. The city cannot cope with the rate of growth and struggles to meet their housing needs. (1 mark)

- Many new residents end up living in self-constructed homes that are vulnerable to damage by weather or fire due to the building materials used. (1 mark)

- Residents live in overcrowded housing – estimates suggest 1 million people live in an area no bigger than 1 mile by 1 mile – (1 mark) without facilities such as water, sanitation and electricity. This leads to the rapid spread of diseases such as cholera. (1 mark)

- Shanty towns are often found close to railway tracks, rivers (which become open sewers) and rubbish dumps, creating dangers for young children and high levels of air and noise pollution for residents. (1 mark)

- Shanty town sites are illegal and residents face the threat of eviction by city authorities at any time. This hinders long-term planning by both residents and local authorities in improving the shanty towns. (1 mark)

- Services – including schools and hospitals – are unable to keep pace with the rapid population growth, meaning that quality of life is poor and many residents struggle to break out of poverty. (1 mark)

- Rising levels of vehicle emissions in Mumbai have led to higher rates of respiratory illnesses amongst residents. (1 mark)

- Residents often rely on the black market to make money. This is often unregulated and involves dangerous jobs in drugs, crime and prostitution. (1 mark)

- Many young children are forced to work to support their families, often collecting waste for recycling at the expense of a proper education. (1 mark)

- The authorities in Mumbai have offered secure long-term legal ownership of land in some areas to encourage residents to improve their housing. This has worked in some areas but the threat of eviction remains high and some shanty towns have been cleared for modern housing and commercial developments. (1 mark)

- Few shanty town residents end up living in these new houses and they often end up on more marginal land. (1 mark)

- Community spirit and strong entrepreneurial links forged over years are broken in this way, for example the recycling district has seen plans for redevelopment, threatening high employment rates in this area. (1 mark)

- Funding from Non-Governmental Organisations (NGOs) and Inter-Governmental Organisations (IGOs), such as the World Bank, has been given to fund the provision of clean water within the shanty towns. These services have struggled to meet the rapid growth in population in Mumbai, although they have improved health levels in small areas. (1 mark)

> ## TOP TIP
> When evaluating impacts of methods used to resolve problems you will need to give a brief outline of the solution. Be careful though as marks are only available for commenting on the effectiveness of these solutions rather than detailing the solutions themselves. It is very important to avoid writing a lot in areas not directly covered by the question.

5. The main method of data collection is the census but candidates may refer to birth and death registrations and border controls. Possible answers might include:

- Data is used to plan and deliver services based on need and demand. For example, information on births allows decisions to be made on investment in maternity care and education services. (1 mark)

- The location of new schools and/or school closures is often linked to long-term projections based on birth rates across the country. (1 mark)

- Information collected via the census allows governments to make decisions and formulate policy on pension levels because they can predict future demand based on responses. (1 mark)

- In some countries population data is used to inform population policies. For example, in China the one-child policy was a response to rapid population growth. (1 mark)

- Migration patterns are recorded by governments to ensure that their criteria are met and that infrastructure is adequate, e.g. housing and service provision in areas where migrants settle. (1 mark)

6. Candidates must refer to impacts on both people **and** the environment. A maximum of 3 marks is available for either part. Answers that make no reference or are over-generalised should not score more than 2 marks.

Possible answers for semi-arid areas might include:

- Land degradation has become a common push factor for the movement of people from the countryside to towns and cities, resulting in the growth of shanty towns. (1 mark)

- Crops have failed and livestock have become weak or died, limiting food supply in affected areas, causing starvation and even death. (1 mark)

- Over-cropping is linked to rapid population growth. This has exhausted nutrient levels in poor quality soils leading to soils that cannot support further crop harvests. (1 mark)

- Soils are left exposed and are vulnerable to wind erosion, e.g. Harmattan, as there are no roots to bind the soil. (1 mark)

- Heavy infrequent rain storms can wash away soil, turning semi-arid areas into deserts, e.g. large areas of the Sahel have been turned to desert in the last few decades. (1 mark)

> **TOP TIP**
> Although no additional marks are available for case study references in this question there is a penalty for candidates who make no reference to areas they have studied. Put the name of your case study at the top of your answer and underline it. Try to bring in place names and/or statistics to support your points.

Possible answers for rainforests might include:

- Land degradation has led to the loss of many important plant and animal species, resulting in endangerment and extinction in some cases. (1 mark)

- Many plants are important for the pharmaceutical industry and loss of these species might limit further research. (1 mark)

- Indigenous people in the Amazon struggle to find prey to supplement slash and burn farming due to the loss of forest. (1 mark)

- Many indigenous people are starting to move to towns and cities, e.g. Rio, increasing pressure on services. Many end up living in shanty towns such as Rocinha. (1 mark)

- As a result of deforestation, nutrient recycling in the soil is reduced and fertile soils are exposed to heavy rainfall and may be washed away. (1 mark)

- Soil and silt often enter the water cycle and can block rivers leading to flooding and loss of fish and marine life. (1 mark)

- Global climate change is linked to forest cover and the rainforest provides an important role in the conversion of carbon dioxide to oxygen. Deforestation increases the greenhouse effect. (1 mark)

SECTION 3: Global Issues

7. (a) Candidates must refer to physical **or** human factors. Where candidates cover both physical **and** human factors both will be marked and the highest individual mark will be awarded. Over-generalised/vague answers that make little reference to named locations will achieve a maximum of half marks. Answers will vary according to the area studied but for the Hoover Dam in the Colorado basin possible responses might include:

Human factors:

- When deciding on a site for the Hoover Dam on the Colorado River, authorities considered access carefully. It was important to ensure construction workers and then maintenance staff could access the dam easily and therefore proximity to settlement was key. (1 mark)

- A nearby market for water and/or electricity limits the prospect of water loss via evaporation if water is being transported over long distances. (1 mark)

- This also reduces the cost of transportation in terms of the amount of infrastructure required to move the water and/or electricity. (1 mark)

- Authorities have to consider the value of the land that will be flooded after the construction of a large dam. Farmland, settlement and important landmarks may be lost, requiring compensation. (1 mark)

- Sites of historical importance, e.g. Native American burial sites, must be treated sensitively to ensure that valuable landmarks are preserved /stakeholders are considered. (1 mark)

Physical factors:

- Geology is a major consideration for planners. It is important that the bedrock is solid due to the weight of the constructed dam and reservoir, so areas of soft/sedimentary rock are avoided. (1 mark)

- Impermeable/non-porous rocks are avoided to prevent water loss through the ground. (1 mark)

- Areas with reliable rainfall and adequate water in their basins are identified to ensure that reservoirs fill to appropriate levels. (1 mark)

- Although many parts of the Colorado basin have semi-arid conditions, planners try to minimise the risk of water loss through evaporation from water stores by avoiding excessively hot locations. Low temperatures are preferred. (1 mark)

- Narrow and deep valleys are preferred as they can be dammed more efficiently and/or at a lower cost. The deep valley provides good storage for the subsequent reservoir. (1 mark)

(b) Possible answers might include:

- The Hoover Dam has provided water for growing desert cities such as Phoenix and Las Vegas, improving quality of life via the increase in landscaped areas and luxuries such as air conditioning and swimming pools. (1 mark)

- Irrigation has allowed agri-business to grow in areas in the south-west of the USA. Many crops can be grown, creating additional and/or more reliable revenue streams in the arid environment. (1 mark)

- Some critics argue that this can lead to water waste as these products could be grown more cheaply and without the need for irrigation elsewhere. (1 mark)

- Irrigation has led to increased salinity in soils, increasing fears that human intervention is degrading the long-term sustainability of the river basin. (1 mark)

- While the river has created a new habitat for over 250 species of bird, some species have vanished from the area since the construction of the Hoover Dam, which flooded habitat and changed the hydrological cycle of the river basin. (1 mark)

- The flooded valley has created an area popular with tourists. Recreation opportunities in Lake Mead, including watersports, have created jobs and provided social benefits. (1 mark)

- Control over the river's discharge has reduced the risk of flood in the river basin. This has avoided expensive repairs as a result of water damage associated with floods. (1 mark)

> **TOP TIP**
> You can mention benefits and adverse consequences in this answer.

8. (a) Answers must show a good understanding of composite indicators. For example, possible answers for the Human Development Index (HDI) might include:

- The HDI is a composite indicator measuring aspects of health, education and economic activity (income linked to purchasing power). This wider approach to measuring development overcomes problems with narrow or single indicators by giving a more accurate picture on development. (1 mark)

- Single indicators often hide wide variations within countries. For example, Saudi Arabia performs well in economic development measures because of the oil industry, however wide inequalities exist between the rich and poor. (1 mark)

- Single economic indicators do not take purchasing power into account. Therefore countries may appear less developed and/or have a lower quality of life as $1 will go a lot further and buy more necessities in some parts of the world. (1 mark)

- Regional variations are masked by single indicators. For example, in Brazil rural and urban areas differ greatly. Rural communities lack services such as health-care and education for youngsters. (1 mark)

- Single indicators are too narrow and do not consider the role of the informal economy. In China subsistence agriculture remains common and is not yet monetized. This appears to devalue the economic output of China. (1 mark)

(b) Candidates must make detailed reference to countries studied to achieve full marks. Possible answers might include:

- The high prevalence of HIV/AIDS in countries in sub-Saharan Africa limits development as public expenditure is diverted in large amounts to treatment of the ill. (1 mark)

- Disease can also limit the economic productivity of the population, lowering GDP and lowering levels of educational attainment as children miss school through illness. (1 mark)

- In small countries it is more cost-effective for authorities to provide services such as health-care and fresh water supplies, e.g. Singapore. Whereas in large nations like China, rural communities are often under-serviced and residents may be unable to access adequate health-care as required, lowering development standards. (1 mark)

- Resource-rich countries are able to export raw materials and use the money to improve living standards through the provision of better services, e.g. effective sanitation in settlements. For example, South Africa is relatively more developed because of the presence of precious minerals such as gold. (1 mark)

- Zimbabwe is relatively underdeveloped due to its political system. In the past, Robert Mugabe has misappropriated aid from foreign governments at the expense of development projects, limiting development in Zimbabwe. (1 mark)

- Foreign countries boycott Zimbabwean produce and refuse to trade with the country in protest at Mugabe's treatment of political opponents. This limits development in comparison with countries who have stronger relations with other countries. (1 mark)

- Climate can influence development. Lesotho is a high, mountainous country with reliable and abundant rainfall. It has a major water management project, regulating water and providing electricity and improving standards of living through heating and lighting in homes. (1 mark)

- Lesotho sells surplus water and electricity, providing an income that can be reinvested in infrastructure, improving development. (1 mark)

TOP TIP

You can write about positive and negative reasons for levels of development in your answer.

9. (a) Answers must refer to named locations. 1 mark can be awarded where candidates refer to **two** specific named examples. Possible answers might include:

- Rising temperatures are linked to melting ice caps in Greenland, the Arctic and Antarctica, leading to a rise in global sea levels. This is also linked to the thermal expansion of water caused by rising sea temperatures. (1 mark)

- Flooding and the risk of flooding and loss of productive agricultural land affects areas like the Maldives and millions living on deltas such as in Bangladesh. (1 mark)

- The growth in climate change refugees may force residents in these areas to seek shelter in neighbouring regions or countries on higher land, putting pressure on resources such as housing, sanitation and water supplies. (1 mark)

- Melting ice caps may alter ocean currents, e.g. the North Atlantic Drift/thermohaline circulation. This current warms the UK and a change may lead to more severe winter weather in the UK. (1 mark)

- More extreme weather leading to drier or warmer conditions will decrease crop yields. In marginal areas, e.g. the Sahel, this will reduce crop yields in areas where food shortages are common and may lead to famine and increase in death rates. (1 mark)

- Conversely, warmer conditions may open up markets for new crops in different parts of the world, e.g. grape harvests may be possible in northern Europe, creating new markets and opportunities to generate money. (1 mark)

- Climate change may improve health by reducing winter-related diseases such as influenza amongst the elderly in the UK. (1 mark)

- However, the 2003 heat wave that affected large parts of Europe was blamed for approximately 40,000 deaths, mainly amongst the elderly. (1 mark)

- Diseases may appear in new regions as the climatic conditions suitable to the spread/ incubation of disease became more extensive due to climate change. For example, the female anopheles mosquito carries malaria but is limited to areas with a warm, humid climate. (1 mark)

- Animal habitat is lost, resulting in the loss of species or movement to marginal areas. This may bring animals into contact with humans – e.g. polar bears entering towns in Canada – resulting in a risk to both animals and people. (1 mark)

TOP TIP

Be careful that you do not write converse/opposite statements as you may not get additional credit. For example, a well-developed sentence about warmer climates preventing crop growth in the Sahel followed by a sentence on warmer climates promoting crop growth in northern Europe will only get one mark. These points would need to have different ideas for 2 marks, e.g. the first point may discuss the impact on nutrition and human health whereas the second point may refer to economic opportunities created by the changing climate.

(b) Possible answers might include:

- Scotland is making progress towards green energy targets. At the end of 2014 it was reported that renewable sources of energy were providing 30% more power and meeting almost 50% of the demand in the country, therefore reducing greenhouse emissions. (1 mark)

- However, renewable energies have been criticised by environmental groups and are often subject to local opposition in planned areas. Wind farms are deemed unsightly and there are concerns about noise pollution: the government have struggled to get public support for development in many proposed sites. (1 mark)

- Renewable energies are more costly than conventional non-renewable sources, discouraging investment as governments and energy providers do not want to increase bills for their customers. (1 mark)

- The rapid growth of populations in the developing world has coincided with rapid development in countries like China and India. The development model/global markets lead these countries to adopt industrial practices leading to the rise of factories and a massive increase in greenhouse emissions. (1 mark)

- As well as approaches to limit greenhouse emissions, developed countries are also investing in strategies to mitigate against the impacts of climate change, e.g. the Thames Flood Barrier. This protects large urban areas but is not feasibly reproduced across the country or affordable for developing nations. (1 mark)

> **TOP TIP**
> Although not worth a huge amount of marks, questions about effectiveness of solutions could also be tackled as a short debate looking at counter arguments about the relative merits of each solution. Again, evidencing knowledge of actual places and the success of strategies will be credited more than generalised answers.

10. (a) Possible answers might include:

- Developing countries often sell raw materials and/or primary products at a low value with resultant lower profits. (1 mark)

- Developing countries often compete with one another and this competition drives down the price for these products. (1 mark)

- Developed nations add value to products through the manufacturing process, leading to higher prices and profits. (1 mark)

- Developed countries place high tariffs/import taxes on processed goods, limiting the opportunity for developing countries to add value to their products. (1 mark)

- These patterns of trade are similar to colonial relationships where developing countries were often forced to buy back processed goods that originated as raw materials in the developing country. (1 mark)

- Trading alliances between the developed countries limit imports through the use of quotas, limiting the volume that developing countries can export, affecting their profits. (1 mark)

- Developing countries may rely on one or two products and have a narrow trading base, e.g. coffee or sugar. This leaves them vulnerable to poor harvests and price shocks caused by over-production. (1 mark)

(b) Possible answers might include:

- Campaigns and Non-Governmental Organisations (NGOs) have been established to encourage governments to cancel debt arrangements, for example, Make Poverty History and Drop the Debt. This allows developing countries to focus on improving their infrastructure (e.g. factories, roads, ports) to promote trade. (1 mark)

- The World Trade Organisation mediates in trade disputes and seeks to promote free trade through the removal of tariffs and quotas. (1 mark). This gives developing nations access to foreign markets, allowing them to earn foreign currency. (1 mark)

- Regional trading alliances have been set up, e.g. The Association of Southeast Asian Nations (ASEAN), allowing these countries to promote trade internally. (1 mark)

- ASEAN countries work together to encourage the area as a global production base by sharing technologies and expertise to spread the benefits of economic growth throughout the region. (1 mark)

> ### TOP TIP
> Don't be afraid to bring in **relevant** knowledge from other subjects. In this question, many of the issues overlap with Modern Studies, and candidates studying Geography in S6 after coming from Modern Studies or candidates studying both subjects in conjunction should know that there are some overlaps in the context of international geo-political issues. This means you will have less to study!

11. (a) Possible answers might include:

- Hydroelectric power is an important energy source in northern Europe, where high rainfall ensures that reservoirs fill quickly, providing water for energy production. (1 mark)

- In the Highlands of Scotland, impermeable rock prevents water loss through seepage into the ground. (1 mark)

- Wind power is effective where there are no barriers to the prevailing wind, which allows regular/uninterrupted wind to power turbines. (1 mark). For example, Whitelee Wind Farm to the south of Glasgow is the largest on-shore wind farm in the UK. (1 mark)

- Geothermal energy is effective in regions that lie on tectonic boundaries. For example, in Iceland, geothermal energy associated with the mid-Atlantic ridge allows heat from the magma to produce steam that is transferred into an energy source. (1 mark)

- Solar energy is found in regions with high levels of sunshine hours to power solar panels, e.g. Spain and Germany. (1 mark)

- Oil is found in the Middle East, where geological conditions allowed organic matter to collect on the then seabed, which over millions of years converted to oil. (1 mark)

- The USA has become a net exporter of gas due to technological developments that have allowed gas to be collected from deep stores within the bedrock using the process of fracking. (1 mark)

- Coal is found in abundance where ancient forests have been buried and were subject to change by high temperatures and compression, creating large stores of carbon. These conditions have occurred in places such as Australia. (1 mark)

(b) A maximum of 2 marks will be awarded for answers that are vague and/or over-generalised. Possible answers might include:

- Non-renewables, such as oil, are effective in meeting energy needs because existing infrastructure (e.g. container ships) is in place to move large quantities around the planet. (1 mark)

- To ensure that energy is cost-effective, developing countries adopt existing technologies to convert non-renewables into energy, allowing them to develop their industrial base and grow economically. (1 mark)

- The use of non-renewables can lead to improvements in the quality of life in parts of the world through the provision of energy to homes, providing light and heat for the first time. (1 mark)

- Burning non-renewables continues to release large quantities of greenhouse gases such as carbon dioxide. These gases trap solar energy in the atmosphere, warming the planet. (1 mark)

- Non-renewables by definition will run out and some experts believe we are reaching the point of peak oil, gas and coal. As resources deplete they will become more expensive, and conflict over these resources may lead to disagreement and even war. (1 mark)

- In many developing countries, wood is often used as a fuel source in homes for heating and cooking as it can be locally sourced and the infrastructure is lacking for getting energy in other ways. As a finite resource this energy source is under pressure from growing populations. (1 mark)

- The clearing of woodland/forest for energy has left areas vulnerable to soil erosion and desertification, e.g. the Sahel. (1 mark)

TOP TIP

Read these questions very carefully. Candidates who rush straight into their answer have been known to confuse similar terms and interpret the question in the wrong way. This question is about **non-renewables** but a candidate in a rush who does not re-read the question may instinctively write about **renewables.** Answers that do not directly respond to the question score no marks regardless of how good the geography is.

SECTION 4: Application of Geographical Skills

12. Candidates must refer to all sources, including the OS map, in their answers. A mark should be awarded each time candidates refer to the resource and offer an explanation with reference to the brief. If candidates make no reference to the OS map then a maximum of 4 marks can be achieved. Grid references and place names should be quoted in the answer. Possible answers might include:

- **An off road course:**
  - The proposed route follows roads less than 4m wide or paths, but runners have to cross the A815 at (122976) and (144854) so will have to cross the road at two points. (1 mark). The route passes along the road for approximately 1km in 1197.

- **Provide a challenging/hilly course:**
  - The west side of the Loch has an undulating path, which may allow for a quick time but does not meet the brief for a challenging course. (1 mark)
  - The route reaches a height of 1000 feet on the east side/second half of the route, providing tough uphill sections for the runners. (1 mark)
  - On the east side of the route, some areas may be exposed to extreme weather conditions and/or strong winds due to the altitude, causing difficulties for runners. (1 mark)

- **Encourages business in the area:**
  - The route starts at Benmore Gardens, where competitors and spectators will spend money in the café and gift shop. (1 mark)
  - Hotels and camp sites (144866) in the surrounding area (Dunoon and Loch Eck side) will see additional business. (1 mark)
  - This is even more valuable in late January when occupancy rates are traditionally low in Scotland. (1 mark)
  - The race is limited to 100 runners, so this limits the amount of additional business being brought into the area. (1 mark)

- **Scenic for competitors:**
  - The route loops around Loch Eck, offering competitors views that may be of interest. (1 mark)
  - However, in large parts the route passes through woodland, which may block loch views around the route. (1 mark)
  - There are few additional sites/landmarks on the route, which may make the route less interesting for some competitors. (1 mark)

- **Provides facilities at the start/finish area:**
  - Car parking is available for runners and spectators at the start and finish area, but there is no marked public transport indicated on the map. (1 mark)

- **Has a suitable transition area (area where runner one finishes and passes on to runner two in the relay teams):**
  - The area around Glenbranter is approximately half-way around the route and has an approach road that could be used to drop off/collect runners. (1 mark)
  - It may become congested in this area if large numbers of runners are completing the event as part of a relay and travelling independently to the village due to the narrow roads and lack of parking. (1 mark)

Mark scheme for Exam C

SECTION 1: Physical Environments

1. Answers should refer to the four main elements in the drainage basin. Possible answers might include:

   - Precipitation is the main input in a drainage basin. Rain clouds are fed by evaporation from water sources including the seas/oceans. (1 mark)

   - The level of water is related to the form of precipitation, the regularity of precipitation as well as the duration and intensity of the precipitation event. (1 mark)

   - Water makes its way into a number of stores within the drainage basin. On the surface, water is stored in lakes and rivers. Water can also be stored on the surface within vegetation, which intercepts precipitation. (1 mark)

   - Water is also stored below the surface as groundwater. The permeability of the soil and rock in the area will determine how much water can be stored as groundwater. (1 mark)

   - Water is transferred on the surface of the basin. This occurs in two ways: as surface run-off where water flows across the surface and via river channels within the drainage basin. (1 mark)

   - Water transfers below the ground through the basin as it makes its way towards the river channel. Throughflow and groundflow refer to the movement of water below the surface and are characterised by slow movement of water. (1 mark)

   - Outputs are the means by which water exits the drainage basin. Water can be lost directly from the channel or water stores through the process of evaporation. (1 mark)

   - Following interception, water can also be lost via trees and vegetation through the process of evaporation from leaves. This is called transpiration. (1 mark)

   - The combined output is known as evapotranspiration. (1 mark)

   - The final output is the loss of water from the basin through streamflow into the sea/ocean as the river ends its journey. (1 mark)

   > **TOP TIP**
   > It is very important to learn and use appropriate terminology. Make revision cards with each key term marked on one side and the definition on the back. You could revise by sorting these into the correct categories or getting someone to test you (by asking for the definition or by reading you the definition and asking you to name the term).

2. No marks will be given for a simple description. Candidates should focus on reasons for the changing rainfall patterns moving inland from the Ivory Coast north towards Mali. Possible answers might include:

- Total rainfall is highest in the south along the coast due to the influence of the Tropical Maritime air mass. The mT air mass is overhead for most of the year and brings warm and wet conditions, increasing annual rainfall. (1 mark)

- At the point where the mT and Tropical Continental (cT) converge, air is heated due to solar energy. Warm air rises and leads to a band of rain known as the Inter-Tropical Convergence Zone. (1 mark)

- The ITCZ, and associated band of rain, tracks north and south over West Africa due to the tilt of the Earth. (1 mark)

- Peak rainfall occurs in Ouagadougou in August, 2 months later than Abidjan as the ITCZ takes time to track north and bring associated rain and the mT air mass. (1 mark)

- Places such as Abidjan in the south have twin peaks of rainfall (in June and November) as the ITCZ tracks north in the summer and returns south in the winter. (1 mark)

- Timbuktu in the north receives very little rain each year as the ITCZ fails to reach or influence this region except for a very short period of time. It is influenced by the cT air mass associated with warm and dry conditions. (1 mark)

- The cT air mass is influenced by the movement of air over the Sahara desert where it warms but fails to collect any moisture. (1 mark)

> **TOP TIP**
>
> Be careful not to get carried away calculating total rainfall etc. in a question of this nature. Marks are awarded for understanding the reasons for variations in rainfall. Including general statements about which place has the highest rainfall and variations in rainfall across the year are as much detail as you require.

3. Candidates must answer both parts of the question. A maximum of 4 marks will be awarded for either part. Answers that are vague and make little reference to case studies will not score as highly as those that do. Answers will vary on the landscape chosen; possible answers for a glaciated landscape such as the Cairngorms might include:

- Traffic congestion is a problem close to tourist attractions such as the Aviemore ski resort, as large numbers of visitors arrive on narrow country roads, inconveniencing local residents and businesses. (1 mark)

- Heavy traffic has caused damage to smaller country roads, resulting in an increase in the number and size of pot holes, which can cause costly repairs for local authorities. (1 mark)

- Visual pollution is an issue: developments such as the Beauly to Denny transmission power line threaten to reduce visitor numbers as tourists complain about the impact on the landscape. (1 mark)

- New wind farms proposed in the National Park cause concern to environmental groups with regard to the damage to wildlife habitat during construction and the risk to rare birds found in the sub-arctic environment once turbines are erected. (1 mark)

- Solutions to travel problems include the promotion via the Cairngorm Mountain website of alternative modes of travel to encourage the use of bus and rail services. 90% of visitors still travel to national parks by car, limiting the impact of such measures. (1 mark)

- Improvements have been made to cycle paths and walking paths to encourage visitors to leave their cars at home, resulting in a reduction of vehicles on summer days. (1 mark)

- Visual pollution has been minimised effectively in a number of ways. For example the funicular railway has been designed with a purple carriage to blend in with the heather bloom. The train also passes through a tunnel to minimise the visual impact. (1 mark)

- Additional attractions in Aviemore have been opened and/or have extended opening hours to develop a wider spread of visitors across the area, minimising congestion. This has worked to an extent but most visitors continue to make their way to the ski resort in the winter, creating congestion. (1 mark)

- Visitor information notices and leaflets are available in accommodations and the Cairngorm Visitor Centre. These have minimised conflict between tourists and farmers in relation to incidents of dropping litter, keeping to footpaths and closing gates. (1 mark)

> ### TOP TIP
> Be careful that you do not repeat yourself in your answers. For example, it would have been easy to mention visual pollution as a conflict associated with both wind farms, the funicular railway and power lines in the case study above. Find a different angle for each conflict in order to maximise marks.

SECTION 2: Human Environments

4. Answers will depend on the landscape chosen. Candidates must refer to a named area within their answer.

   Possible answers for a named rainforest area might include:

   - Land in the Amazon Rainforest has been purchased by conservation groups/NGOs such as WWF. They prohibit deforestation and replant indigenous trees to maintain the ecosystem. (1 mark)

   - Afforestation schemes have been used. Replanting of trees provides roots that bind the soil, preventing erosion by water or wind. (1 mark)

   - The Amazon Region Protected Areas (ARPAs) – now in its second phase – is a scheme between local, national and international organisations and has protected an area the size of California from deforestation. (1 mark)

   - Charities also work with indigenous people to ensure that slash and burn and/or shifting cultivation farming continues to be practiced. This method is environmentally sensitive and land is allowed to lay fallow to recover/minimise degradation. (1 mark)

- Agro-forestry schemes seek to promote sustainable development within the rainforest to ensure that employment and economic gains are balanced with protection of the ecosystem. The tree canopy protects the soil from the damaging impact of heavy rain showers. (1 mark)

- Trees and crops are often planted in the same plots and crops are rotated. This helps maintain and protect the soil by minimising the depletion of nutrients. (1 mark)

- Education about the consequences of deforestation plays an important role in ensuring those who develop and manage the forest are able to make decisions based on sustainable principles. (1 mark)

Possible answers for a named semi-arid area might include:

- Stone lines can be placed at the bottom of slopes and when overland/surface flow moves sediment it is trapped. Locals then rake this material back over the source area, minimising land degradation. (1 mark)

- Small-scale schemes to improve irrigation bring greater water security to areas, ensuring that vegetation does not die. This encourages natural vegetation to grow. Native species can also be reintroduced to increase the amount of vegetation. (1 mark)

- Large-scale dams are also constructed in some areas to ensure regular water supply but these are extremely expensive and can have negative environmental consequences. (1 mark)

- Crop rotation ensures that the same nutrients are not always taken from the soil. This allows the soil to recover and the cultivation of various crops that harvest at different times maintains vegetation cover. (1 mark)

- Larger trees are often planted in lines perpendicular to the prevailing wind. This prevents the wind from blowing away the nutrient rich top soil, protecting against desertification. (1 mark)

- Drought-resistant shrubs and plants are planted. These succeed in areas with limited access to water/low rainfall. The roots of the plants bind the soil and prevent erosion by heavy rain. (1 mark)

- Controlling grazing by spreading around feeding and water stations prevents animal trampling on the ground, which can compact the soil and lead to higher rates of erosion. (1 mark)

5. Candidates must refer to both the economic impacts **and** the impacts on the citizens of Chad based on the pyramids showing a rapidly growing population. Possible answers might include:

- Chad will struggle to meet financial obligations in relation to the growing demand for childcare services such as midwives and education, removing finance from other priority areas of development. (1 mark)

- Many children may not receive an education, lowering literacy rates and hindering development in countries with rapidly growing populations. (1 mark)

- A rapidly growing population will put pressure on resources such as housing. Many residents may end up living in shanty towns with poor levels of sanitation, leading to rapid spread of disease. (1 mark)

- Population may outstrip resources and Chad may face issues relating to land degradation if land is over-cultivated or not left fallow to recover nutrients. (1 mark)

- As children reach adulthood, Chad will struggle to provide enough employment, leading to unemployment and poverty. Families may not be able to afford essentials such as food and water. (1 mark)

- However, Chad may be able to use the young population to attract inward investment, leading to job prospects and investment in infrastructure, benefiting all residents of Chad. (1 mark)

6. Possible answers might include:

- In Glasgow many of the worst tenement houses were demolished and replaced with high-rise flats, which allowed high population densities to be rehomed in areas such as the Gorbals. (1 mark)

- High-rise flats were equipped with indoor bathrooms and heating, leading to health benefits for residents of these areas. (1 mark)

- Many residents of overcrowded inner city areas were accommodated in nearby towns such as Barrhead and Paisley. This was a quicker solution than rebuilding due to the existing infrastructure in these areas. (1 mark)

- More recently, high-rise flats have been demolished in areas including Govan. These flats have had significant problems linked to the loss of community spirit, damp and the breakdown of lifts, inconveniencing residents. (1 mark)

- Modern tenement-style apartments have been built within the inner city, e.g. the Crown Street Regeneration Project. Services such as libraries and leisure centres are included in new developments to encourage longer-term success. (1 mark)

- Loans and part ownership schemes have been launched to help existing residents afford houses in redeveloped inner city areas. This is in response to criticism that many existing residents are often priced out of new developments. (1 mark)

SECTION 3: Global Issues

7. (a) Answers will depend on the case study chosen. For example, possible answers for the Nile River might include:

- The climate in the lower basin is characteristic of a hot desert with low rates of precipitation below 250 mm per annum. The rain that falls is unpredictable, resulting in seasonal flow in the river. (1 mark)

- High temperatures throughout the basin lead to rapid water loss via evaporation, reducing the discharge/flow in the lower basin in Egypt. (1 mark)

- Water from various sources (including Lake Victoria) combine to form the White Nile which joins the Blue Nile, with its source in the Ethiopian Highlands. Both combine to feed the river. (1 mark)

- Lake Victoria is particularly vulnerable to change in rainfall and evaporation rates, causing fluctuations in levels of discharge in the Nile. (1 mark)

- In Sudan, rainfall creates a flooded landscape for half the year, while the dry season leaves the landscape desert-like in appearance for the other half, resulting in poor soils. (1 mark)

(b) Candidates must cover both benefits **and** adverse consequences to achieve full marks. Answers that are over-generalised and make little reference to named locations should get a maximum of 4 marks. Possible answers might include:

- Dams allow for greater control of river discharge, allowing authorities to monitor the river and minimise flood risk in the drainage basin. This reduces the risk of damage to property, crops and habitat. (1 mark)

- In areas with water-control projects, the guarantee of clean water provision can provide health benefits through a reduction in diseases, such as cholera, caused by communities relying on contaminated water. (1 mark)

- However, in warmer climates the creation of stagnant reservoirs can increase incidences of diseases such as schistosomiasis. (1 mark)

- In many water projects, the creation of a reservoir has led to recreation opportunities and associated social benefits because people can take part in water sports and have scenic areas to visit. (1 mark)

- Hydroelectric power generated by water projects often leads to greater levels of foreign investment/development of industry-creating jobs, e.g. in Borneo following the construction of the Bakun Dam. (1 mark)

- Water projects can create new habitats for bird species but the flooded valley often damages existing habitat. (1 mark)

- This can lead to a reduction in tourism associated with the previous landscapes/ species endemic to the affected area. (1 mark)

- Reservoirs may flood large areas of forest cover, reducing the capacity for trees to recycle carbon dioxide into oxygen. This has an impact on global warming. (1 mark)

8. Candidates may choose to answer parts (a) and (b) separately or together. A maximum of 6 marks will be awarded for either section.

(a) Possible answers for malaria might include:

- Life expectancy is lowered and death rates are higher in affected areas. Malaria kills over 1 million people each year, with over 90% of these deaths occurring in Africa. (1 mark)

- Countries have to spend money on treating malarial patients, education programmes and malarial prevention approaches and research into the disease, diverting funds from other important development goals such as investment in education. (1 mark)

- Roll Back Malaria reports that some of the most affected countries spend up to 40% of their health budget on malaria. (1 mark)

- Those infected have prolonged absence from work and/or school, impacting on levels of attainment and economic output, hindering development. (1 mark)

- Unicef reports that Africa as a continent would have a GDP $12billion per year higher if malaria was eradicated. (1 mark)

- In some areas, communities can harvest less than 50% of their crops because much of the labour required is unavailable due to malaria. This reduces personal income and food provision in vulnerable communities. (1 mark)

- Families often have to invest heavily in treatment and take time off work to care for loved ones or help them travel to treatment centres; therefore the impact on yield and/or school attendance is multiplied beyond the patient. (1 mark)

- Many TNCs are put off investing in malarial areas due to poor health records, and tourism as an industry struggles because of the reputation and risks associated with the disease. This inhibits growth. (1 mark)

(b) Possible answers might include:

- Although no vaccine is available, organisations such as the Bill and Melinda Gates foundation are currently researching this. Drugs like quinine, chloroquinine, larium and malarone have been developed in an attempt to kill the parasite. (1 mark)

- Drugs based on the Chinese herb Artemisia have been effective in killing off the parasite. (1 mark)

- Attempts to kill the adult mosquito population include spraying areas with insecticides such as DDT. (1 mark)

- Trials have taken place where mosquitoes are genetically modified to produce sterile mosquitoes or killer mosquitoes, in order to reduce the spread of the insect and therefore malaria. (1 mark)

- Sleeping under bed nets treated with insecticide in vulnerable areas stops the mosquito from biting people and infecting them when they are asleep. (1 mark)

- The addition of larvae-eating fish to stagnant water helps to reduce the mosquito population because fish such as the muddy loach clear the area. (1 mark)

- Planting eucalyptus trees next to areas of stagnant water can remove breeding grounds as the tree is very absorbent, removing areas where mosquitoes can breed successfully. (1 mark)

> **TOP TIP**
> In (b) it is very important that you say how each method actually helps to control the disease you have selected. You cannot list solutions and expect to score multiple marks. Therefore plan to write a sentence about each solution.
> If you choose to write about malaria, it is helpful to organise your solutions under headings, e.g.:
> * Strategies to manage the plasmodium
> * Strategies to manage the adult mosquito population
> * Strategies to manage the mosquito larvae

9. (a) 1 mark can be awarded where candidates refer to **two** specific named examples (ocean currents, animal species, data etc.) within answers. Possible answers might include:

* Changes in global temperatures may be linked to monsoon and drought conditions in landmasses with a Pacific coast, as a result of El Niño and La Niña events. (1 mark)

* Warming temperatures may lead to bleaching in coral reefs, affecting the species that this important ecosystem supports. (1 mark)

* In locations such as the Great Barrier Reef, this may affect tourism revenues and associated jobs. (1 mark)

* Some areas may become new tourist hotspots if they become drier and warmer. For example, countries such as the UK may see more stay-at-home tourism if the school holidays become drier, resulting in economic growth. (1 mark)

* Tropical disease, such as yellow fever, may extend into new areas as climates warm, resulting in the spread of deadly diseases. (1 mark)

* Sea levels are rising as a result of melting glaciers and polar ice caps. Warmer climates also lead to an expansion of the water in the oceans, further adding to this problem. (1 mark)

* The melting ice caps may provide quicker and more effective transport routes between North America and Europe due to the opening of new sea routes, resulting in greater trade opportunities. (1 mark)

* People living on low-lying flood plains and river deltas are at risk of damage to their properties, and loss of businesses and farmland, and may have to relocate. (1 mark)

> **TOP TIP**
> 1 mark is available for two named case studies in this question. This is an additional mark so if you mention, for example, the threat to 10% of land species including the polar bear and the melting of ice caps in areas such as Antarctica, you will get 1 mark. This is the limit, but it is good practice to include examples as often as possible.

(b) Possible answers might include:

* Renewable approaches to meeting energy needs is a key strategy in reducing greenhouse emissions. In Scotland, hydroelectric power is helping to meet almost 50% of the country's energy needs through renewable sources. (1 mark)

- However, new schemes face resistance due to the impact on the visual landscape, and/or the impact on the drainage basin and habitat for wildlife, making it difficult to meet targets to obtain the majority of energy from renewables. (1 mark)

- International agreements have been criticised because some powerful countries (e.g. the USA) did not sign up to the Kyoto Protocol. This limited the impact of the agreement because many signatories felt they did not have to abide to agreements and some countries such as Canada subsequently withdrew from the agreement. (1 mark)

- Countries who remained in the agreement often broke their carbon emission targets but limited sanctions and/or appetite to punish this has held back the ambitious targets to cut emissions laid out at the outset. (1 mark).

- Travel is one of the biggest contributors to greenhouse emissions and authorities have tried to encourage a shift to walking, cycling, car-pooling and greater use of public transport. The relative falling cost of the automobile and the rise in car ownership in countries such as India and China has limited the impact of these approaches on a global scale. (1 mark)

- In developed countries the counter-urbanisation movement means that the car has become an even more important part of life in the last decade, although investments in cycle lanes/public bike hiring schemes have improved participation rates in cities such as London. (1 mark)

> **TOP TIP**
> You must refer to **at least two** strategies in this answer. Make sure you do this. The inclusion of '**at least two**' indicates that the examiner is looking for variety and detail about each of your solutions, so focus on two or three at the most in this situation.

10. (a) Candidates must refer to the impact of trade on both the people **and** environment of **developing** nations to achieve full marks. Possible answers might include:

- In developing countries the production of primary products is one of the key features of international trade. Associated low labour costs mean that workers are unable to pay for medical bills, food etc. for their families. (1 mark)

- In many areas, children aged 15 and below work to ensure that their families have enough money for basics such as food, denying them access to education. (1 mark)

- Furthermore, working environments can be unsafe and conditions unfair, leading some to feel that children are being exploited and put at risk in order that products are available for export to developed countries. (1 mark)

- The production of palm oil in Indonesia provides foreign currency, which can be used to pay off debt and invest in infrastructure and services, including the provision of hospitals and medical centres. (1 mark)

- Indonesian rainforests are home to 10–15% of all known species of plants, mammals and birds, and deforestation for palm oil and paper production is threatening these species through the destruction of their habitat by forcing them onto marginal lands. (1 mark)

- Fires are used to clear large areas of rainforest, releasing huge volumes of carbon dioxide into the atmosphere. Greenhouse gases are linked to changes in the climate. (1 mark) affecting those living in developing countries through the increase in frequency and strength of extreme weather including floods, droughts and tropical storms. (1 mark)

> **TOP TIP**
>
> Make sure you only write about impacts on developing nations, as specified in the question. Any reference to impacts on developed nations will receive no credit.

(b) Possible answers might include:

- Fair trade promotes a fair price and wage for workers in developing countries, allowing them to improve their quality of life. Fair trade involves 1.4 million workers, mainly in Africa and the Middle East, benefiting them by providing additional funds that can be used for education, home improvements and health-care costs. (1 mark)

- Fair trade guarantees prices and allows these workers/small communities to plan for the medium-to-long-term, rather than responding to short-term shocks in the price of their commodities. This has allowed communities to invest in infrastructure, including the provision of clean water and sewage pipes. (1 mark)

- The US government has paid countries in Central and South America to preserve areas of rainforest. This ensures that these nations do not lose out financially by not developing these areas for mineral extraction or cattle ranching. (1 mark)

- This has been successful in some areas by reducing rates of deforestation, however the rules also prevent small-scale/sustainable projects run by indigenous groups in the rainforest, inhibiting their quality of life. (1 mark)

- Pressure to cancel debt has led to the abolition of repayments in many contexts, allowing governments to reinvest this money in programmes such as universal education through to the age of 15. However, in many countries debt has risen and the global total is still increasing. (1 mark)

11. (a) No marks will be available for just describing the growth in demand (i.e. energy demand is growing worldwide but at a faster rate in developing countries). Candidates must account for this growth. Candidates must refer to factors in developing **and** developed countries in their answer to achieve full marks. Answers that fail to consider both types of country should be marked out of 4. Possible answers might include:

- The growth in demand for energy is linked to economic growth and population growth. In many developing countries, the population is increasing dramatically, leading to an increase in demand for energy. (1 mark)

- Authorities in India have committed to providing energy in rural communities. The roll-out of this programme has dramatically increased the demand for energy in a country of over 1 billion people. (1 mark)

- Much of this demand comes from increasing connections to the mains grid for electricity for lighting, heating, air conditioning and other domestic appliances to enhance quality of life. (1 mark)

- The demand for luxury goods that use high amounts of energy continues to increase in developed countries as standards of living increase, e.g. washing machines and dishwashers are now commonly found in many homes. (1 mark)

> **TOP TIP**
> Be careful that ideas for developed countries and developing countries are not too similar. You will not get additional credit for repeating the same idea in a different geographic context. The above examples are very similar and the developed country example may not score an additional mark unless new ideas are introduced.

- As developing countries develop, disposable income rises, leading to a growth in goods such as cars. The number of cars in developing countries is growing rapidly, leading to a growth in energy use. (1 mark). This could also be true of developed countries. (1 mark)

- In developed countries, air travel is growing dramatically and internal flights are now extremely common. (1 mark)

- The movement of goods around the world – particularly of manufactured goods from developing countries to developed nations – increases energy use by large container ships. (1 mark)

- Rapid industrialisation in parts of Asia and Africa has led to a proliferation of factories, leading to a surge in energy demand. (1 mark)

- In developed and developing nations, new technologies are increasing energy efficiency in the industrial process, reducing the demand for energy. (1 mark)

(b) Candidates must refer to suitability of renewable approaches. 1 mark can be awarded where **two** specific named relevant examples are given. Possible answers might include:

- Solar energy has been successful in countries with above-average hours of sunshine. Where the sun is more intense (e.g. Spain and Greece) and cloud cover is limited, solar panels can be powered successfully. (1 mark)

- Wave power is effective where the fetch (the distance the wave has travelled) is greatest, giving the most powerful waves to convert to energy. The south-west coast of the UK (e.g. Cornwall) is well placed to generate energy in this way. (1 mark)

- Wind power is producing a large share of energy in Germany where sites without shelter provide regular/uninterrupted wind to move turbines. (1 mark)

- Concerns have been raised in Scotland that this method is not cost-effective compared to non-renewable approaches to generating energy. (1 mark)

- Hydroelectric power is ideally suited to areas with steep relief and reliable rainfall. The Highlands, and hanging valleys of Scotland, can be easily dammed and provide ideal sites. (1 mark)

- The drop in relief to the valley floor provides the drop required to allow the water to gain enough speed to power turbines in HEP stations. (1 mark)

TOP TIP

Try to ensure you give examples to get the additional case study mark. A specific named example should **not** just be the name of a country. For example, when discussing wind farms a good example would be Whitelee Wind Farm south of Glasgow, the largest on-shore wind farm in Europe.

SECTION 4: Application of Geographical Skills

12. (a) Possible answers might include:

- Good access by road from East Kilbride and close to a number of other large settlements, providing easy access for potential customers. (1 mark)

- The site is on the edge of East Kilbride providing space for growth and a cheaper location than a more central site. (1 mark)

- The site is close to other facitlities e.g. the golf course at 633564 and sports centre at 637566 which may bring increased custom to the new attraction as potential customers may plan a day out to the area to visit multiple attractions. (1 mark)

- The land is not too steep but undulating so construction on this site would not be straightforward and would have required leveling of the land adding to the cost of the project. (1 mark)

- The site is not particularly close to nearby rivers and the elevated site offers protection from the risk of flooding. (1 mark)

(b) Possible answers might include:

- The opening of the new attraction and associated national publicity may bring additional custom for other local attractions and businesses, creating a multiplier effect. (1 mark)

- The opening of a large attraction will create jobs in the local area – both during the construction phase and for staff now that the attraction is up and running. (1 mark)

- The site is near to a castle (627561) and heritage park and the modern development may spoil the natural and ancient landscape in this particular part of the town through visual pollution. (1 mark)

- Calderglen Country Park is one of the most visited attractions in this part of Scotland and the new attraction may reduce the number of visitors and affect businesses in this part of East Kilbride. (1 mark)

- Local residents may be disturbed by increased traffic in the area, which may bring noise pollution and cause difficulties in completing short journeys. (1 mark)

- Traffic travelling from the south may have to pass through the centre of the town, leading to congestion in the busy CBD. (1 mark)

- Potential loss of green belt land at the edge of East Kilbride may endanger wildlife habitat and threaten some species. (1 mark)

Mark scheme for Exam D

SECTION 1: Physical Environments

1. Possible answers might include:

   - Climate – gleys are found in cool regions such as the Tundra. These conditions mean natural vegetation is limited, providing little nutrients and litter, reducing the development of gley soils. (1 mark)

   - Vegetation breaks down very slowly in the cold climate, resulting in an acidic humus. (1 mark)

   - Drainage – gley soils can be peaty and are characterised by waterlogging, which denies the oxygen that supports the presence of soil organisms. (1 mark)

   - Relief – gley soils tend to be found in flat landscapes, resulting in limited surface run-off and further adding to the problem of poor drainage and waterlogging. (1 mark)

   - Soil organisms extract limited oxygen supplies from the soil, decolouring the soil and leading to a grey/blue appearance. (1 mark)

   - Parent material is often found in the B horizon of the soil due to frost shattering of the parent material. (1 mark)

   - Soil heave brings this weathered material up through the soil. At times this material may breach the surface. (1 mark)

2. Candidates must refer to conditions and processes to achieve full marks. Well-annotated diagrams that have well-developed explanations of conditions and processes can achieve full marks. Answers will depend on feature chosen. For example, possible answers for terminal moraine might include:

   - Moraine is the term given to material transported and deposited by a glacier. The terminal moraine marks the furthest extent of the ice, and material is deposited as the ice starts to melt. (1 mark)

   - This process moves material from the zone of accumulation through the glacier to the point of melting/zone of ablation. (1 mark)

   - Material is added to the terminal moraine by the process of bulldozing, where material is pushed down the valley by the glacier. (1 mark)

   - The terminal moraine is made up of unsorted rocks, boulders, stones, pebbles etc. as the glacier has the energy to move material of all different sizes. (1 mark)

   - As the glacier retreats, the terminal moraine is left behind and can form a dam, creating a ribbon lake in the newly created U-shaped valley. (1 mark)

> ### *TOP TIP*
> There is less to say about some features. In this instance there is not a huge amount of information about the formation of features of glacial deposition. If asked to do two features, a good second choice might have been drumlin as there are opposing ideas about formation, thereby giving more to discuss. If you choose outwash plain, you should be talking about fluvio-glacial (melt water) deposits.

3. Possible answers might include:

- Reflection from the atmosphere reduces the amount of energy reaching the Earth's surface (26%) as solar energy reflects off clouds and gas and dust particles. (1 mark)

- Reflection levels vary across the planet and according to weather conditions. Higher levels of cloud cover lead to greater rates of energy loss. (1 mark)

- The reflective properties of the Earth's surface (albedo) add to energy loss. This is greatest at the white surfaces, including the polar ice caps. (1 mark)

- At the equator, where forest cover dominates, the dark colour is less reflective and absorbs energy, increasing the % of solar energy absorbed in these areas. (1 mark)

- Energy is also lost through absorption by the atmosphere. Clouds, gas and dust absorb solar energy (approximately 20%) within the atmosphere and store this, reducing the energy that can be absorbed by the Earth's surface. (1 mark)

4. Candidates must explain strategies **and** comment on their effectiveness. Answers that only cover one part should be marked out of 5. Vague and/or over-generalised answers that make little or no reference to places should not be awarded more than 3 marks.

> ### *TOP TIP*
> You cannot simply describe the strategies designed to improve shanty towns. You must **explain** these. This also provides opportunity for development of answers, which will get 2 marks. For example the answer, 'Roads have been improved and tarred in areas such as Rocinha,' is only description. You must develop these ideas:
> 'Roads have been improved and tarred in areas such as Rocinha. This means that residents are able to move through the shanty town and travel further for work. (1 mark)
> As a result of new roads, trade with the wider urban market in Rio is improved, allowing residents in the shanty town to export goods for sale.' (1 mark)
> There are a range of joining phrases that force us to give more detail/explain when writing answers:
> - … this means that …
> - … therefore …
> - … as a result …
> If you get feedback from your teacher that you are not giving enough detail and/or explaining when asked, add the following to the end of your sentences. They force you write more.

Possible answers might include:

- The authorities in Rio de Janeiro have administered self-help schemes in favelas. Residents are given materials to enhance their homes, such as breeze blocks, tiles and glass. This reduces the risk of weather-related property damage or fires in wooden properties. (1 mark)

- Through the engagement with locals/residents in the shanty town, money is saved on labour, allowing investment in other areas. (1 mark)

- Houses in the Rocinha shanty town have been connected to water and electrical supplies along with sewage pipes, improving the quality of life for residents. This means that the risk of disease spreading in an overcrowded area is reduced. (1 mark)

- Site and service schemes provide a range of basic amenities and allow residents to buy or rent the land their home sits on. This removes the threat of eviction for residents in the shanty town. (1 mark)

- This encourages further self-improvement and longer-term planning by residents and has been successful in Rocinha. (1 mark)

- Investment in services such as education and health in rural areas aims to reduce in-migration to the city and relieve pressure on shanty towns by removing push factors. (1 mark)

- Although partially successful, large numbers of young migrants leave the countryside every day in the hope of a better lifestyle in the city, therefore hindering the process of improving the shanty towns. (1 mark).

- Some shanty towns have been cleared and replaced with high-rise apartments but this has not been entirely successful as many of the poorest residents are unable to afford the rent and end up living in more peripheral areas. (1 mark)

5. Vague and/or over-generalised answers should achieve no more than 2 marks. Possible answer might include:

- Many Polish people have moved to the UK in search of employment. Unemployment rates in the last 10 years have been consistently higher in Poland, and therefore many young Poles migrate in search of work. (1 mark)

- Average wages are higher in the UK than Poland. Poles often leave skilled jobs in Poland to work in the service sector in the UK, where they can earn more money in jobs that require lower levels of qualification. (1 mark)

- When Poland joined the EU in 2004, Poles were able to move freely to the UK. Many moved for quality of life decisions, including amenities, services and bright lights such as entertainment facilities in large cities such as London. (1 mark)

- There is a well-established Polish community in the UK, which makes transition into life in a new country easier and provides informal support to migrants. (1 mark)

TOP TIP

Know what is happening today with your case studies. This will impress the examiner and shows that you are not relying solely on your class notes. For example, at the time of writing migration from Poland to the UK is slowing as wages in Poland rise in relative terms. Changing exchange rates and an improving economy have resulted in fewer Poles migrating to the UK and a number choosing to return home. By reading news websites and/or following the headlines for the duration of your course, you will be amazed how many stories are connected to unit content.

6. Possible answers might include:

- The cost of preparing, printing and administering the census is prohibitive in developing countries, where public health and housing are more immediate concerns. (1 mark)

- In countries such as Burkina Faso, literacy rates are low. This means additional support is required or forms are incomplete/partially completed. (1 mark)

- Data collected may not be accurate for various reasons. For example, under-registration is common in China where the one-child policy impacts on registration of new births. (1 mark)

- Countries with large numbers of people living in temporary accommodation (e.g. refugees in Syria) or shanty towns (e.g. Dharavi in India) have limited records to administer the census, making accurate collection of data difficult. (1 mark)

- Geographic isolation and poor infrastructure make it difficult to administer the census in remote parts of large countries, such as the Amazon Rainforest in Brazil. (1 mark)

> **TOP TIP**
>
> In this question type it is very easy to slip into a list-style response. Make sure you connect points directly to the question. For example, how does each factor you are discussing relate to the collection of census data? A lot of marks are lost due to poor technique.

SECTION 3: Global Issues

7. (a) A maximum of 2 marks should be awarded for answers that are vague and/or over-generalised. Possible answers might include:

- A reliable and constant supply of water from tributaries, rainfall and snowmelt is a crucial consideration to ensure that the project has enough supply to meet the demand. (1 mark)

- Areas with lower temperatures are preferred because evaporation rates from the increased surface area associated with reservoirs can lead to the loss of water in warm regions. (1 mark)

- Tectonically active areas (near plate boundaries) are avoided because damage of infrastructure could lead to catastrophic damage to the water project and nearby settlements if the dam was to fracture or fail. (1 mark)

- Narrow and deep valleys are easier and cheaper to dam than wide/shallow valleys and can effectively store large volumes of water. (1 mark)

- The biodiversity of the river basin must be considered because it is important that valuable habitat for endangered species is maintained. Species may be rehomed or other sites may have to be considered. (1 mark)

(b) Candidates must refer to socio-economic **and** environmental impacts of a **named** water management project. 1 mark can be awarded where candidates refer to **two** specific named examples *within* the case study area. Possible answers for the Nile River might include:

Socio-economic impact:

- The supply of food in the region is more secure, leading to a healthier population/ reduction in poor health associated with a lack of food. (1 mark)

- Lake Nasser provides opportunities for recreation (e.g. fishing for tigerfish), providing jobs and boosting the economy of the area. (1 mark)

- Approximately 100,000 Nubians were forced to relocate because their land was flooded following the construction of the Aswan Dam and the reservoir it created. (1 mark)

- The generation of electricity from hydroelectric power has resulted in an improved infrastructure and attracted a number of domestic and foreign-owned industries to establish bases in the area. (1 mark)

- Egypt is deeply in debt to Russia, who financed the construction on the Aswan Dam (total cost $1 billion). (1 mark)

Environmental impact:

- Water levels can be controlled, providing a reliable discharge in the river that can be matched to the needs of the plants and animals who depend on the Nile. (1 mark)

- The river has become more saline due to an increase in rates of evaporation, affecting crop yields and the success of different crops in the region. (1 mark)

- Poor irrigation associated with the project has caused waterlogging of many soils, resulting in a loss in the nutrient value of previously high yielding farmland. (1 mark)

8. (a) Possible answers might include:

- Natural resources present in some countries, e.g. oil in Saudi Arabia, generate wealth that can be used to improve public services, which is harder to achieve in resource-poor nations such as Ethiopia. (1 mark)

- Conflict diverts resources away from schools and hospitals to the purchase of weapons and/or training of armed forces. In Afghanistan, this has made development difficult in comparison to relatively stable countries such as India. (1 mark)

- Conflict destabilises the area, limiting long-term planning and discouraging investment into the country. (1 mark)

- Natural harbours and coasts make trade more viable in countries such as Singapore than in land-locked Chad. This generates foreign income, which can be used to purchase imports to improve quality of life. (1 mark)

- Countries like Haiti and Bangladesh are regularly affected by natural disasters that cause damage to vital infrastructure. Repairs use up resources and make it harder for these nations to develop. (1 mark)

- Countries often end up in debt as they borrow money to pay to rebuild their schools and hospitals and end up locked into agreements that remove the power to make independent decisions on their own development. (1 mark)

- The climate in sub-Saharan Africa limits agricultural production and makes it very difficult to survive. A surplus to be sold is hard to produce, leading to a subsistence country with limited funds to improve basic infrastructure including waste disposal and water supply. (1 mark)

(b) Possible answers might include:

- Barefoot doctors – locals who are trained to treat simple/common injuries and illnesses – are highly effective as they are trusted and can travel to remote areas with limited medical facilities. (1 mark)

- Preventative measures reduce overall treatment costs. Treatment costs are much higher than preventative vaccinations programmes. (1 mark)

- This is very important in developing countries with limited finance for public services, where diseases like polio and measles reduce attendance at work and school, further hindering development. (1 mark)

- Education about the role of bed nets and covering skin at dusk is important in preventing people from contracting malaria. A bed net costs £5 and these are often distributed by international charities. (1 mark)

- Messages about health-care are often shared through talks in rural villages or on posters. These measures are accessible by those who are illiterate, increasing their audience and impact. (1 mark)

- Local workers are often involved with charities in the construction of local clean water projects/toilet facilities. Materials are provided, and training/transferable skills encourage wider usage and acceptance in the region. (1 mark)

> **TOP TIP**
> You must develop points of evaluation for a mark to be given. For example, you cannot rely on saying it is effective because it is cheap. You must say it is cheap, making it affordable for developing countries. Cost will only be credited once, so make sure you have a range of evaluative ideas ready. These could be positive or negative. For some ideas you may be able to give both a positive and a negative response.

9. (a) Possible answers linked to the increase in greenhouse gases might include:

- The increase in the world's population is linked strongly to rising levels of methane, a by-product of rice cultivation in Asian countries. (1 mark)

- The increase in demand for beef has also seen levels of methane rise dramatically from belching cattle. (1 mark)

- CFCs are released when refrigerators and air conditioning systems are not disposed of correctly. CFCs trap heat in the atmosphere, warming the planet. (1 mark)

- Nitrous oxides are released from vehicle exhausts. The level of gas in the atmosphere has risen dramatically as car ownership continues to grow in developed nations and is rising exponentially in developing nations such as China and India. (1 mark)

- Carbon dioxide is released during the burning of forests, which are being cleared to create land for palm oil cultivation in Indonesia, reducing the ability of the forest to convert carbon dioxide into safe oxygen. (1 mark)

- The burning of fossil fuels to generate electricity and power vehicles around the globe continues to grow, releasing more carbon dioxide into the atmosphere. (1 mark)

(b) Possible answer might include:

- Warming of the oceans causes expansion in the water, resulting in rising sea levels. (1 mark)

- This is compounded by additional water from melting ice caps and glaciers. (1 mark)

- Melting ice may place some areas that rely on annual snow/ice melt in water crisis. (1 mark). For example, the River Chillon in Peru supports 2 million people and is fed by glacial meltwater from the Andes, but the ice coverage has reduced by 40% in recent decades. (1 mark)

- Low-lying coastal areas, e.g. millions of residents living on the delta in the Bay of Bengal in Bangladesh, are at risk of flooding and damage to property. (1 mark)

- Longer growing seasons may be possible in Scandinavia and the UK as a result of global warming, leading to an increase in yields and income for farmers. (1 mark)

- Longer and more severe wild-fire seasons in places such as western USA and parts of Australia are due to rises in temperatures and falling levels of precipitation. (1 mark)

- Severe weather episodes including drought, heat waves and hurricanes have become more severe in magnitude and more frequent, for example major typhoons in the Philippines in recent years. (1 mark)

> ### TOP TIP
> This question asks you to refer to countries you have studied. Simply naming countries will not get you marks but if you bring in relevant examples by referring to named events, dates or places to support your point then you could achieve additional credit for making a development of your point. See for example the marking instruction that refers to Peru and glacial meltwater in the answer above. This answer brings in two place names and two statistics, earning additional credit. Learning your case studies is key to achieving a pass in Geography; focus on a few key facts in each topic area, and write these down and get someone to test you on them.

10. (a) Possible answers might include:

- Developed countries predominantly sell processed/manufactured goods that have value added. These sell for more than raw materials, generating larger profits. (1 mark)

- Developing nations are often over-reliant on one primary export product. Poor harvests as a result of pests or climatic conditions can leave a country vulnerable and widen the trade gap because there are no other products to fall back on as a safety net. (1 mark)

- Competition from other nations producing the same raw materials keeps the price of primary products low, lowering income. (1 mark)

- Primary products are subject to price fluctuations or may be set by trading on global exchange markets. These conditions are often more favourable to developed nations as consumers than developing nations as producers. (1 mark)

- Developed nations limit the access that developing nations have to their markets through the use of tariffs and quotas, which make imports relatively more expensive or limit the quantity of import allowed in. This hinders developing nations. (1 mark)

(b) Candidates must explain strategies **and** comment on their relative success to achieve full marks. 1 mark can be awarded where candidates refer to **two** specific named examples *within* the case study area. Possible answers might include:

- Trade organisations such as the Free Trade Zone in Africa aim to reduce trade inequalities by promoting trade within the group of member nations. (1 mark)

- This has given Africa a stronger/more unified voice in trade negotiations with developed countries, resulting in improvements in trade relations such as the removal of some tariffs. (1 mark)

- This move has been supported by the World Trade Organisation (WTO) as it seeks to remove barriers to international trade. However, wealthy nations hold great influence/positions of power in major Inter-Governmental Organisations (IGOs), preventing wholesale changes that may adversely impact on developed nations. (1 mark)

- Member countries are encouraged to purchase raw materials and share technology and expertise to promote regional development. Preventing the outflow of money for goods boosts growth in the region, creating jobs and promoting further investment in export industries. (1 mark)

- Fair trade promotes a fair price and wage for workers in developing countries, allowing them to improve their quality of life. Fair trade has added a premium of over €1,000 million for workers, providing funds for education of children, home improvements and health-care costs. (1 mark)

- Fair trade guarantees prices and allows member communities and/or individuals to make longer-term plans. This has given confidence for communities to invest in infrastructure, including the provision of clean water and sewage pipes. (1 mark)

- The Ecuadorian government recently accepted money from foreign governments and private individuals to limit oil extraction within the rainforest. This approach has protected environmentally important areas while allowing Ecuador to maintain a balance of trade. (1 mark)

- Some experts believe measures like this are unsustainable moving forward as we near peak oil and prices inevitably rise, making it more attractive for Ecuador to begin exporting from the rainforest again. (1 mark)

11. (a) Possible answers might include:

- China is the world's largest exporter of coal. Coal is found in abundance where ancient forests have been buried and subjected to change by high temperatures and compression, creating large stores of carbon. (1 mark)

- Russia is the world's leading exporter of natural gas. Gas is found in Russia due to geological processes where decaying plant and animal matter collected, often at the bottom of the ocean, and has been converted into gas stores. (1 mark)

- Hydroelectric power is important in areas where reliable rainfall can be stored in reservoirs and used to generate electricity. (1 mark)

- Underlying geology (impermeable rock) is required to ensure that stored water does not escape from the store. (1 mark)

- Solar energy has developed in areas with high numbers of sunshine hours, such as Spain. (1 mark)

- Israel has no oil and poor political relationships with oil-rich neighbours, so solar energy has become price comparable. Panels are placed in desert regions. (1 mark)

- Although physical factors are extremely important, human factors have relevance in the global distribution of energy sources. Investment by government and/or levels of research and development by industry and universities is important in the location of renewable energy. For example, wind energy in Scotland and solar technology in Germany. (1 mark)

(b) 1 mark can be awarded where candidates refer to **two** specific named relevant examples.

Possible answer might include:

- Biofuels provide energy through the burning of plant matter. In developing countries, indoor air pollution has led to increased levels of respiratory illness and causes of death in countries that rely on biofuels. (1 mark)

- Biofuels are often grown as a cash crop for domestic commercial use or export. They are found on the most fertile lands, reducing crop yields. (1 mark)

- Some indigenous forest dwellers have been forced from their land and/or animal habitats have been lost due to clearing of forest for biofuels. (1 mark)

- Biofuels do not release huge stores of carbon dioxide, so they are more environmentally friendly than burning fossil fuels. (1 mark)

- However burning/clearing large forests to grow biofuels removes the ability of trees to recycle carbon dioxide into oxygen. (1 mark)

- Wind power has been very effective in areas where wind is reliable and the prevailing wind is uninterrupted by natural or man-made barriers. (1 mark)

- One drawback is the creation of energy on days when wind is calm and turbines are generating less energy. Technology at the moment is not efficient at storing surplus energy generated by turbines. (1 mark)

- The high cost of construction is prohibitive for developing countries although the energy produced is cheaper in the long run. (1 mark)

SECTION 4: Application of Geographic Skills

12. Candidates must refer to all sources including the OS map in their answers. A mark should be awarded each time candidates refer to the resource and offer an explanation with reference to the brief. If candidates make no reference to the OS map, a maximum of 4 marks can be achieved. Possible answers might include:

- **Be spectator friendly and accessible at various points:**
  - Spectators can access the route by public transport in the centre of Glasgow, e.g. Queen Street Station and Central Station are 1–1.5km from the start/finish area. (1 mark)
  - The Glasgow Subway underground trains allow spectators access at various points on the course such as Kelvinbridge in the West End. (1 mark)
  - The Subway could be used to access more than one part of the course. (1 mark)
  - There are few car parks indicated on the map and road closures would make it difficult to see the event by car. (1 mark)

- **Pass important landmarks to promote the city:**
  - The start/finish area follows the River Clyde, providing views of the water. (1 mark)
  - The route passes the University, museums and several parks, which will stand out on television pictures. (1 mark)
  - The route could be improved if it went closer to historic buildings, e.g. the Cathedral. (1 mark)

- **Be suitable for all participants:**
  - The transect shows that there are some short uphill climbs in the city centre section of the course, which may cause difficulties for some cyclists. (1 mark)
  - The lap set-up will allow participants to choose a distance suitable to them, with good access via public transport back into the city centre for those who only want to complete a short distance. (1 mark)

- **Start and finish at a suitable location:**
  - Glasgow Green is close to public transport but parking may be difficult as it is on the edge of the CBD. A start/finish area towards the inner city or suburbs may provide more space. (1 mark)